AF706983

EVOLUTION THEORY: HISTORICAL BACKGROUND AND FUNDAMENTALS

EVOLUTION THEORY: HISTORICAL BACKGROUND AND FUNDAMENTALS

Prof. Dr. Vinay Kumar Singh

Delve PUBLISHING
www.delvepublishing.com

Evolution Theory: Historical Background and Fundamentals

Prof. Dr. Vinay Kumar Singh

Delve Publishing
224 Shoreacres Road
Burlington, ON L7L 2H2
Canada
www.delvepublishing.com
Email: orders@arclereducation.com

© 2024 Delve Publishing

ISBN: 978-1-77469-949-2 (Hardcover)

This book contains information obtained from highly regarded resources. Reprinted material sources are indicated and copyright remains with the original owners. Copyright for images and other graphics remains with the original owners as indicated. A Wide variety of references are listed. Reasonable efforts have been made to publish reliable data. Authors or Editors or Publishers are not responsible for the accuracy of the information in the published chapters or consequences of their use. The publisher assumes no responsibility for any damage or grievance to the persons or property arising out of the use of any materials, instructions, methods or thoughts in the book. The authors or editors and the publisher have attempted to trace the copyright holders of all material reproduced in this publication and apologize to copyright holders if permission has not been obtained. If any copyright holder has not been acknowledged, please write to us so we may rectify.

Notice: Registered trademark of products or corporate names are used only for explanation and identification without intent of infringement.

Delve Publishing publishes wide variety of books and eBooks. For more information about Delve Publishing and its products, visit our website at www.delvepublishing.com.

ABOUT THE AUTHOR

Professor Vinay Kumar Singh, Ph. D., Department of Zoology, DDU Gorakhpur University, born on 15th July, 1969, has a brilliant academic carrier. Prof Singh has involved himself in research on Fasciolosis and its snail vectors control since 1993. Current research interests are Isolation, characterization and biological evaluation of Natural Plant Products and pest control, Vector-Borne Diseases, Toxicology & Pharmacology. Prof Singh has University teaching and research experiences of more than 30 years. Prof Singh has successfully isolated and identified pure molluscicidal compounds from more than a dozen medicinal plants and observed their mode of action on the snails at cellular level. His work is recognized internationally in the field of formulation and use of synthetic and bio-pesticides for integrated pest/ vector-diseases management. Prof Singh has devised a bioassay method for pesticides using isolated organs of snails. Currently, he is working on the pesticidal property of the different formulations of cow urine of Indian cow breeds singly or synergistically with bioactive product of plant molluscicides. The research publication on fasciolosis control by means of characterized plant derived pesticides, synergistic combinations, bait formulations or by reducing the reproductive capacity of vector/host to limit the population below threshold level which are widely quoted and acclaimed world over. Prof Singh has published 154 research papers in 94 leading International Journals of repute.

TABLE OF CONTENTS

LIST OF FIGURES

LIST OF ABBREVIATIONS

AA	Amino Acid
CNV	Copy Number Variation
CRISPR	Clustered Regularly Interspaced Short Palindromic Repeats
DNA	Deoxyribonucleic Acid
EVO-DEVO	Evolutionary Developmental Biology
GWAS	Genome-Wide Association Study
HGT	Horizontal Gene Transfer
LBA	Long Branch Attraction
LUCA	Last Universal Common Ancestor
MGE	Mobile Genetic Element
MRCA	Most Recent Common Ancestor
MSA	Multiple Sequence Alignment
OTU	Operational Taxonomic Unit
PAM	Point Accepted Mutation
PAUP	Phylogenetic Analysis Using Parsimony
PCR	Polymerase Chain Reaction
PE	Punctuated Equilibrium
RFLP	Restriction Fragment Length Polymorphism
RNA	Ribonucleic Acid
ROS	Reactive Oxygen Species
SNP	Single Nucleotide Polymorphism
STEM	Science, Technology, Engineering, and Mathematics
STEME	Science, Technology, Engineering, Mathematics, and Evolutionary Biology

PREFACE

The concept of evolution stands as one of the most captivating and enduring narratives in the realm of science. It is the story of life's ceaseless transformation, a tale that unfolds over millions of years. At its core, evolution is the process by which species change and adapt to their environments, resulting in the rich tapestry of biodiversity we see today. Within the expansive narrative of evolution, various theories and perspectives have emerged, each contributing its unique thread to this intricate fabric.

Charles Darwin, the name synonymous with evolution, introduced the theory of natural selection in his seminal work, "On the Origin of Species." Natural selection posits that the environment acts as a selective force, favoring individuals with advantageous traits, leading to their increased survival and reproduction. Over time, these traits become more common in the population, driving species' adaptation to their surroundings.

However, the story of evolution doesn't end with Darwin. The Modern Synthesis, also known as Neo-Darwinism, integrated genetics into evolutionary theory. This synthesis melded the principles of Mendelian genetics with Darwin's ideas, providing a comprehensive framework for understanding how genetic variation arises and is passed on from generation to generation.

Beyond the confines of Neo-Darwinism, the extended evolutionary synthesis emerged. This perspective recognizes that evolution is not solely driven by natural selection and genetic mutation. It encompasses epigenetics, developmental biology (Evo-Devo), and other mechanisms that influence how organisms develop and adapt. This broader view acknowledges the complexity of evolutionary processes and the interplay of multiple factors.

To truly appreciate the narrative of evolution, one must also delve into the rich fossil record. Fossils provide a glimpse into the past, offering snapshots of ancient life forms that once roamed the Earth. They serve as a critical source of evidence, bridging the gap between the present and the distant past. Transitional fossils, in particular, provide a compelling testament to the gradual changes that have occurred over eons.

Molecular evidence further bolsters the case for evolution. By comparing the DNA and protein sequences of different species, scientists have unveiled the genetic relationships between organisms. This molecular tree of life not only supports the concept of a common ancestor but also reveals the intricate branching patterns that have given rise to the vast diversity of life forms we see today.

The story of evolution also extends into the realms of ecology, biogeography, and behavioral biology. It explores the intricate relationships between organisms and their environments, the geographic distribution of species, and the behavioral adaptations that facilitate survival and reproduction. These facets underscore the holistic nature of evolution and its connection to every aspect of life.

The study of evolution is not confined to the realm of biology alone. It has profound implications for medicine, agriculture, conservation, and even our understanding of human nature. Evolutionary medicine, for instance, explores how our evolutionary history has shaped our susceptibility to diseases and informs strategies for healthcare. In agriculture, knowledge of evolution plays a crucial role in breeding crops and livestock for improved yields and resistance to pests.

Yet, despite the wealth of evidence and the explanatory power of evolutionary theory, it remains a subject of debate in some quarters. Creationism and intelligent design offer alternative narratives to the diversity of life, sparking controversies in educational curricula and public discourse. However, the overwhelming consensus among scientists is that evolution is a well-established scientific theory supported by an extensive body of evidence.

So, the story of evolution is a grand narrative that transcends the boundaries of time and disciplines. It is a tapestry woven with threads of genetics, paleontology, molecular biology, and ecology. It is a story that has the power to illuminate our understanding of life's origins, the diversity of species, and the interconnectedness of all living things. In unraveling this tale, we gain insights into the past, present, and future of life on Earth. Evolution, in all its complexity, stands as a testament to the remarkable journey of life itself.

This book titled, "Evolution Theory: Historical Background and Fundamentals" is a comprehensive and informative publication on the subject of evolution. This book provides readers with a deep understanding of the historical development of evolutionary theory and its fundamental principles. It covers key concepts, notable scientists, and the evolution of ideas about how species change over time. This book is valuable for anyone interested in the history and foundational principles of evolutionary biology.

CHAPTER 1

Introduction to Evolution

CONTENTS

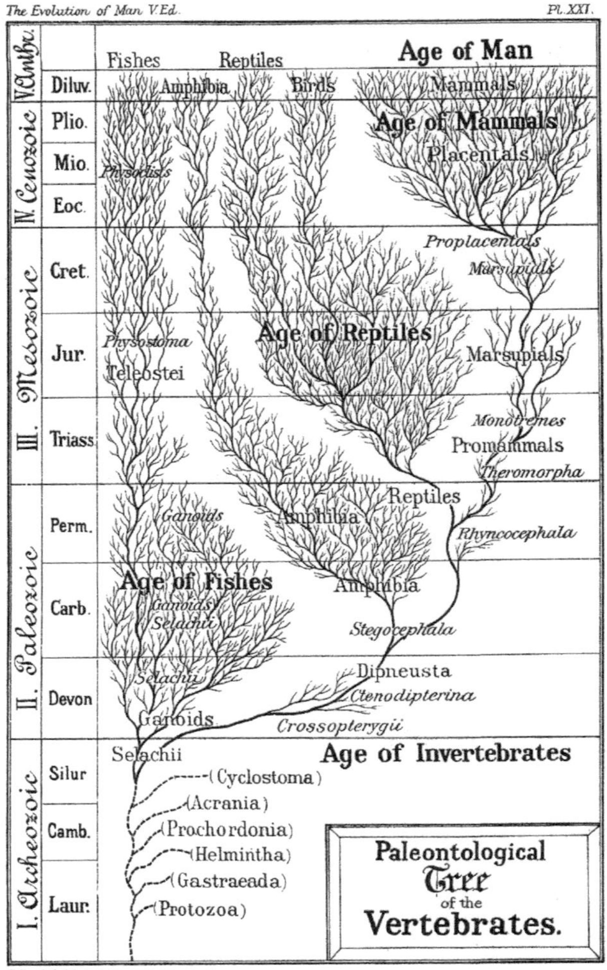

Figure 1.1. Introduction to evolution.

Source: *By Ernst Haeckel – Erst Haeckel, Public Domain, https://commons.wikimedia.org/w/index.php?curid=4246342; Creative Commons License.*

1.1. THE CONCEPT OF EVOLUTION

The concept of evolution stands as one of the most profound and transformative ideas in the history of science. It is a concept that delves deep into the very heart of life itself, unraveling the intricate tapestry of biodiversity that adorns our planet. At its core, evolution is a concept that encapsulates the notion of change, adaptation, and the inexorable passage of time as it shapes the living world.

1. **Descent with Modification:** At the heart of the concept of evolution lies the fundamental idea of "descent with modification." This concept posits that all living organisms share a common ancestor. This means that every species on Earth, from the tiniest microbe to the mightiest whale, is part of a grand family tree with its roots in the distant past. As these ancestral life forms diversified and adapted to different environments, they underwent modifications over countless generations. This continuous process of change and divergence gave rise to the remarkable diversity of life we witness today.
2. **Natural Selection:** A cornerstone of the concept of evolution is the elegant mechanism of natural selection, introduced to the world by Charles Darwin in the 19th century. Natural selection acts as the sculptor of evolution, refining species over time. It operates on the principle that within any population of organisms, there exists variation in traits. Some of these traits provide individuals with a better chance of surviving and reproducing in their specific environment. These advantageous traits are passed on to the next generation more frequently, gradually altering the characteristics of the population. Over time, this process can result in the adaptation of a species to its environment, a process known as "survival of the fittest."
3. **Genetic Variation:** For evolution to occur, genetic variation is paramount. Genetic variation arises through several mechanisms, including mutation, recombination, and gene flow. Mutations are spontaneous changes in an organism's DNA, introducing new genetic material into a population. Recombination occurs during reproduction when genes from two parents combine, producing offspring with unique combinations of traits. Gene flow, or the movement of genes between populations, can also introduce new genetic diversity.
4. **Speciation:** Evolutionary processes can lead to the formation of new species, a phenomenon known as speciation. When populations of a species become isolated from one another—whether through geographic barriers, ecological niches, or other

factors—they can accumulate genetic differences over time. If these differences become significant enough, the isolated populations may no longer be able to interbreed and produce fertile offspring. This reproductive isolation marks the emergence of two distinct species from a common ancestor.

5. **Fossil Record:** The fossil record serves as a tangible and compelling testament to the concept of evolution. Fossils are the preserved remnants of ancient life forms, providing glimpses into the history of species on Earth. They reveal the existence of species long extinct and offer a snapshot of transitional forms that bridge the gap between modern organisms and their ancient ancestors. Fossil evidence helps us trace the evolutionary journeys of various groups of organisms.
6. **Molecular Evidence:** In the modern era, advances in molecular biology have provided powerful evidence for evolution. Scientists can compare the DNA and protein sequences of different species, revealing genetic relationships and common ancestry. The molecular tree of life offers a fascinating glimpse into the genetic connections that underpin the diverse web of life on Earth.
7. **The Broader Implications of Evolution:** The concept of evolution extends far beyond the confines of biology. It has profound implications for fields such as medicine, agriculture, and conservation. In evolutionary medicine, the historical legacy of human evolution informs our understanding of disease susceptibility and provides insights into how our bodies have adapted to past challenges. In agriculture, knowledge of evolution guides breeding programs to enhance crop yields and resistance to pests. Conservation biology leverages evolutionary principles to protect and manage biodiversity.
8. **Controversies and Ongoing Debate:** While the concept of evolution is widely accepted within the scientific community, it has not been without controversy. Alternatives such as creationism and intelligent design propose different narratives for the diversity of life, sparking debates in educational curricula and public discourse. Nevertheless, the overwhelming consensus among scientists is that evolution is a well-established scientific theory supported by an extensive body of evidence.

In essence, the concept of evolution is a grand narrative that transcends the boundaries of time and disciplines. It is a story that weaves together genetics, paleontology, molecular biology, ecology, and more. It is a tale that sheds light on the origins of life, the myriad forms it has taken, and the intricate web of relationships among all living beings. Evolution, with its elegance and complexity, stands as a testament to the remarkable journey of life itself—a journey that continues to unfold, offering new discoveries and deeper insights into the mysteries of existence.

1.1.1. Descent with Modification

At the core of evolutionary theory, "descent with modification" is a fundamental concept that encapsulates the idea that all living organisms are related through a common ancestry, and over time, species change and diversify through a process of modification from their ancestors. This concept underpins the entire framework of evolutionary biology and helps explain the remarkable diversity of life on Earth (Figure 1.2).

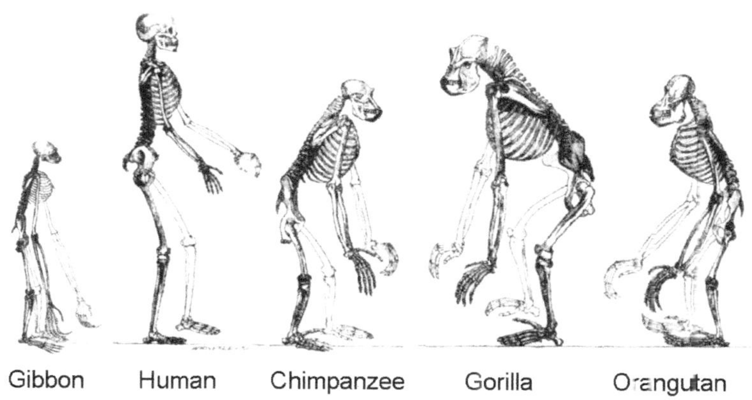

Figure 1.2. Descent with modification.

Source: *By the original uploader was TimVickers at English Wikipedia – Transferred from: en.wikipedia to Commons, Public Domain, https://commons.wikimedia.org/w/index.php?curid=2785168; Creative Commons License.*

1.1.1.1. Key Elements of Descent with Modification

1. **Common Ancestry:** Descent with modification posits that all living organisms share a common ancestor. This means that no

matter how different two species may appear today, they can trace their lineage back to a single, ancient life form that lived in the distant past. The common ancestor represents the point at which the evolutionary journey began.

2. **Gradual Change:** According to this concept, species do not change suddenly or dramatically. Instead, changes accumulate over vast spans of time, often imperceptibly from one generation to the next. These gradual changes may involve alterations in physical traits, behaviors, or genetic makeup.
3. **Divergence:** As descendants of a common ancestor reproduce and spread into different environments, they undergo modifications that lead to divergence. Over time, these modifications accumulate, resulting in the development of new characteristics and the emergence of distinct lineages or species. Divergence is the key driver of biodiversity.
4. **Adaptation:** As species diverge and evolve, they often adapt to the specific environments in which they live. This adaptation occurs through a process known as natural selection. Individuals with traits that enhance their survival and reproduction in a particular environment are more likely to pass on those traits to their offspring. Over generations, this leads to the prevalence of traits that are advantageous in a given ecological niche.
5. **Fossil Evidence:** The fossil record provides tangible evidence of descent with modification. Fossils are the preserved remains or traces of ancient organisms, and they document the existence of species that lived in the past. Transitional fossils, in particular, serve as compelling examples of organisms that display characteristics intermediate between their ancestors and their descendants.
6. **Molecular Evidence:** Advances in molecular biology have provided additional support for the concept of descent with modification. By comparing the DNA and protein sequences of different species, scientists can uncover genetic relationships and identify common genetic ancestors. This molecular evidence corroborates the patterns of descent and diversification observed in the fossil record.

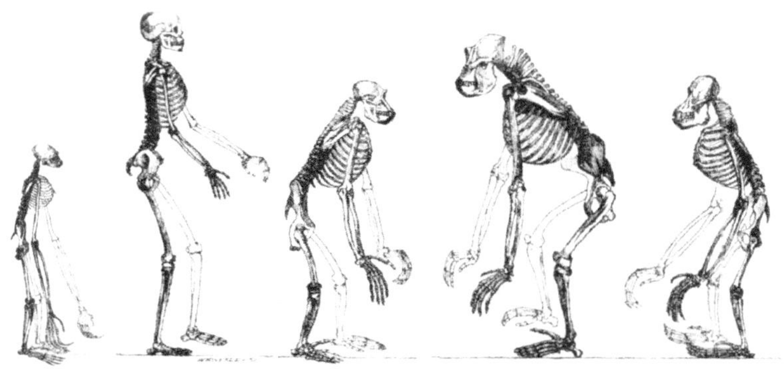

1.1.1.2. Implications of Descent with Modification

1. **The Tree of Life:** Descent with modification is often depicted as a branching tree, with the common ancestor at the base and the branches representing the various lineages of species that have evolved over time. This metaphorical "tree of life" reflects the shared heritage and evolutionary relationships among all living organisms.
2. **Understanding Biodiversity:** This concept provides a comprehensive framework for understanding the diversity of life on Earth. It explains how species have adapted to various ecological niches, leading to the immense variety of organisms we see today.
3. **Biogeography:** Descent with modification also helps explain the distribution of species around the world. Species that share a common ancestry are often found in regions with similar environmental conditions, reflecting their historical connections.
4. **Medical and Agricultural Applications:** Knowledge of descent with modification is essential in fields like medicine and agriculture. Understanding how organisms have evolved can inform medical research, as well as the breeding of crops and livestock for desirable traits.
5. **Conservation Biology:** Conservation efforts often take into account the evolutionary history of species. Protecting species with unique evolutionary lineages helps preserve biodiversity and the ecological roles that different species plays.

In summary, the concept of "descent with modification" is a foundational idea in evolutionary biology. It provides a framework for understanding how all living organisms are interconnected through common ancestry and how they have diversified and adapted to their environments over millions of years. This concept, supported by a wealth of evidence from fields such as paleontology and molecular biology, continues to be a cornerstone of modern science.

1.1.2. Natural Selection

Natural selection is one of the central principles of evolutionary biology. It's a process by which populations of organisms change over time as a result of differential survival and reproduction of individuals with certain heritable traits. This mechanism, proposed by Charles Darwin and Alfred Russel Wallace in the mid-19th century, is fundamental to our understanding of how species adapt to their environments and evolve over generations (Figure 1.3).

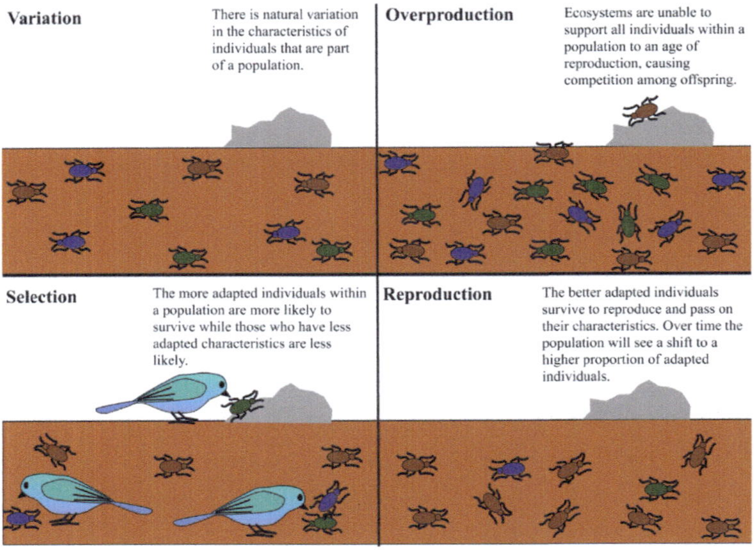

Figure 1.3. Natural selection.

Source: *https://en.wikipedia.org/wiki/File:Natural_Selection.png; Creative Commons License.*

1.1.2.1. Key Components of Natural Selection

1. **Variation:** Within any population of organisms, there is genetic variation. This variation arises due to genetic mutations, recombination during reproduction, and gene flow from other populations. This diversity results in differences among individuals in traits like size, color, shape, behavior, and physiology.
2. **Heritability:** The traits that exhibit variation must be heritable, meaning they can be passed down from one generation to the next through genetic inheritance. If a trait is not heritable, changes in that trait within a population will not persist over time.
3. **Differential Survival:** Natural environments often present challenges and limited resources. Not all individuals within a population will survive to reproduce. Some individuals will have

traits that give them a better chance of surviving and reproducing in their specific environment, while others may be less successful. This leads to differential survival rates among individuals with different traits.

4. **Differential Reproduction:** Individuals with advantageous traits not only survive better but also tend to produce more offspring than those with less advantageous traits. As a result, the traits that confer a survival advantage become more common in the next generation.
5. **Environmental Pressure:** Natural selection operates in response to environmental pressures, which can include predation, competition for resources, changes in climate, or other factors that influence an organism's ability to survive and reproduce. Traits that enhance an organism's fitness (i.e., its ability to survive and reproduce) in the face of these pressures become more prevalent.

1.1.2.2. Illustration of Natural Selection

Let's consider an example to illustrate natural selection:

- Imagine a population of giraffes with varying neck lengths. Some giraffes have longer necks, while others have shorter necks. In their environment, there is limited access to leaves high in trees, which serve as their primary food source. Giraffes with longer necks can reach higher leaves more easily, while giraffes with shorter necks struggle to do so.
- Variation: The population exhibits genetic variation in neck length, with some giraffes having longer necks and others having shorter necks.
- Heritability: Neck length is a heritable trait, as it's influenced by genetic factors.
- Differential Survival: Giraffes with longer necks have a survival advantage because they can access more food. They are less likely to go hungry or starve.
- Differential Reproduction: Giraffes with longer necks, having more food, are also healthier and more likely to reproduce successfully.
- Environmental Pressure: The limited availability of high leaves in trees exerts environmental pressure favoring longer necks.

- Over time, as longer-necked giraffes produce more offspring, and those offspring inherit the genetic predisposition for longer necks, the proportion of giraffes with longer necks in the population increases. This is an example of natural selection in action, as the population gradually adapts to its environment through the spread of advantageous traits.

1.1.2.3. Consequences of Natural Selection

Natural selection is the primary mechanism by which species adapt to their environments. It can lead to the formation of new species, the development of complex adaptations, and the diversity of life on Earth. Over geological time scales, natural selection has played a crucial role in shaping the characteristics and behaviors of organisms across a wide range of ecosystems.

So, natural selection is a foundational concept in evolutionary biology, explaining how the fit survive and reproduce, passing on their advantageous traits to the next generation. It is a dynamic process that continually shapes the biological diversity we observe in the natural world.

1.1.3. Genetic Variation

Genetic variation refers to the diversity of genetic material within a population of organisms. It is a fundamental concept in biology and genetics, as it underlies many biological processes, including evolution, adaptation, and the inheritance of traits (Figure 1.4).

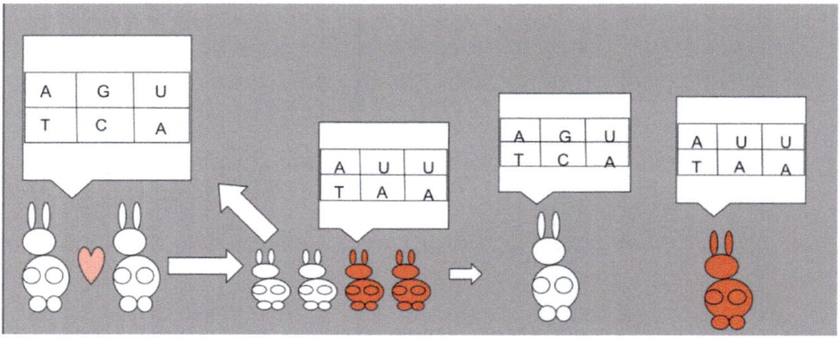

Figure 1.4. Genetic variation.

Source: *By SaulU123 – Own work, CC BY-SA 4.0, https://commons.wikimedia.org/w/index.php?curid=69569253; Creative Commons License.*

1.1.3.1. Key Aspects of Genetic Variation

Origins of genetic variation:

1. **Mutation:** These are random changes in an organism's DNA sequence. They can occur spontaneously during DNA replication, as a result of exposure to environmental factors (e.g., radiation, chemicals), or due to errors in DNA repair processes. Mutations introduce new genetic material and can lead to the creation of new alleles (variants of genes) or changes in existing alleles.
2. **Recombination:** These occurs during sexual reproduction when genetic material from two parents is combined to produce offspring. This process shuffles and recombines the genetic information, creating unique combinations of alleles in the offspring. This generates genetic diversity within a population.
3. **Gene Flow:** Or migration, happens when individuals from one population move and interbreed with individuals from another population. This movement of genes between populations introduces new genetic variation and can homogenize genetic differences between populations.

1.1.3.2. Types of Genetic Variation

1. **Single Nucleotide Polymorphisms (SNPs)**: These are the most common type of genetic variation and involve a change in a single DNA base pair. They can affect traits and susceptibility to diseases.
2. **Insertions and Deletions (Indels):** These involve the addition (insertion) or removal (deletion) of a small section of DNA. They can cause frame shifts in genes, altering protein production.
3. **Copy Number Variations (CNVs)**: These refer to variations in the number of copies of a particular DNA segment. They can have significant effects on gene dosage and gene expression.
4. **Structural Variations:** These involve larger-scale alterations in DNA, such as inversions, translocations, or duplications. These can impact gene function and genome structure.

1.1.3.3. Significance of Genetic Variation

1. **Evolution:** Genetic variation is the raw material for evolution. It provides the genetic diversity upon which natural selection,

genetic drift, and other evolutionary forces act. Populations with greater genetic variation are more adaptable to changing environments.

2. **Adaptation:** Genetic variation allows a population to adapt to new or changing environmental conditions. Natural selection acts on advantageous genetic variants, leading to the spread of traits that enhance an organism's fitness in its environment.
3. **Disease Susceptibility:** Genetic variation plays a role in an individual's susceptibility to diseases. Some genetic variants may increase the risk of certain diseases, while others may provide protection.
4. **Biodiversity:** Genetic variation contributes to the diversity of species and ecosystems. It is a key driver of speciation and the development of unique traits in different populations and species.

1.1.3.4. Measurement of Genetic Variation

1. **Genetic Markers:** Scientists use genetic markers like SNPs to study genetic variation. These markers are specific DNA sequences known to vary among individuals and populations.
2. **Genome Sequencing:** Advances in DNA sequencing technology have made it possible to sequence entire genomes, allowing researchers to identify genetic variations at a comprehensive level.

1.1.3.5. Maintenance and Change of Genetic Variation

The level of genetic variation within a population can change over time due to various factors, including natural selection, genetic drift, migration, and mutation. Natural selection tends to increase the frequency of advantageous alleles while reducing the frequency of disadvantageous ones. Genetic drift can cause random changes in allele frequencies, particularly in small populations. Migration can introduce new alleles, increasing genetic diversity. Mutation continues to generate new genetic variation.

Thus, genetic variation is the cornerstone of biological diversity and evolution. It arises from mutations, recombination, and gene flow and serves as the raw material upon which evolutionary processes act. Understanding genetic variation is essential for studying the genetics of traits, disease susceptibility, and the evolution of species.

1.1.4. Speciation

Speciation is the process by which one species splits into two or more distinct species. It is a fundamental concept in evolutionary biology and explains how the immense diversity of life on Earth has arisen. Speciation occurs when populations of the same species become reproductively isolated from one another, preventing interbreeding, and leading to the independent evolution of each group (Figure 1.5).

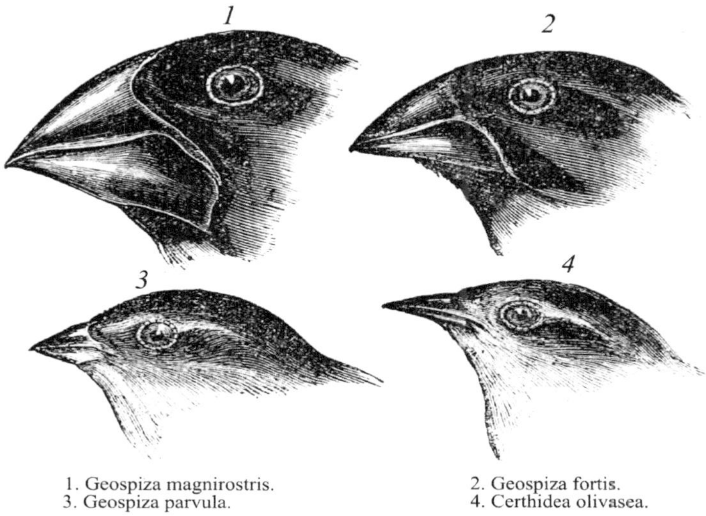

Figure 1.5. Speciation.

Source: *By John Gould (1804–1881) – From "Voyage of the Beagle" as found on http://darwin-online.org.uk/converted/published/1845_Beagle_F14/1845_Beagle_F14_fig07.jpg]; also, online through Biodiversity Heritage Library at https://www.biodiversitylibrary.org/page/2010582., Public Domain, https://commons.wikimedia.org/w/index.php?curid=3918303/; Creative Commons License.*

1.1.4.1. Key Elements of Speciation

1. **Isolation:** The first step in speciation is the isolation of populations. This isolation can be geographic, ecological, temporal, or behavioral in nature. It prevents individuals from one population from interbreeding with individuals from another population.

2. **Geographic Isolation:** This occurs when physical barriers like mountains, rivers, or oceans separate populations. Over time, these populations adapt to their distinct environments, leading to divergence.
3. **Ecological Isolation:** Populations may occupy different ecological niches within the same geographic area, reducing their opportunities to interbreed. For example, one group might feed on different types of food or occupy different habitats.
4. **Temporal Isolation:** Populations may have different breeding seasons or times of day when they are active, reducing the chances of interbreeding.
5. **Behavioral Isolation:** Differences in mating behaviors, courtship rituals, or communication can lead to reproductive isolation. Individuals from one population may not recognize or respond to the cues of individuals from another population.
6. **Genetic Divergence:** Once isolated, populations can begin to accumulate genetic differences over time. This divergence may result from mutation, genetic drift (random changes in allele frequencies), or natural selection acting on different environments or selective pressures in each population.
7. **Reproductive Isolation:** As genetic differences accumulate; reproductive barriers can develop. These barriers prevent individuals from different populations from successfully mating and producing viable, fertile offspring when they do encounter each other.
8. **Prezygotic Barriers:** These are barriers that prevent the formation of a zygote (fertilized egg) and include mechanisms like differences in mating behaviors, mating times, or physical incompatibilities.
9. **Postzygotic Barriers:** These are barriers that occur after the formation of a zygote and include mechanisms like hybrid inviability (hybrid offspring do not survive to maturity), hybrid sterility (hybrid offspring are sterile), or reduced hybrid fitness (hybrids have lower fitness compared to non-hybrids).
10. **Completion of Speciation:** Once reproductive isolation is complete, the two populations are considered separate species. They may continue to evolve independently, adapting to their respective environments and accumulating further genetic differences.

1.1.4.2. Types of Speciation

1. **Allopatric Speciation:** It occurs when populations become geographically isolated from one another. Over time, the isolated populations adapt to their unique environments and diverge genetically. This is one of the most common forms of speciation.
2. **Sympatric Speciation:** It occurs within the same geographic area. It is often associated with the development of reproductive barriers due to ecological or behavioral factors. For example, if two populations of a species exploit different food sources within the same habitat, they may evolve into distinct species over time.
3. **Parapatric Speciation:** It occurs when populations are adjacent to each other but have limited gene flow due to partial isolation. This can result from gradients in environmental conditions, leading to adaptation to different niches along the gradient.

1.1.4.3. Significance of Speciation

1. **Biodiversity:** Speciation is a major driver of biodiversity, leading to the vast array of species on Earth.
2. **Adaptation:** Speciation allows populations to adapt to specific ecological niches, leading to increased fitness and survival in their respective environments.
3. **Evolutionary History:** The history of life on Earth is a record of speciation events. Understanding these events helps us unravel the evolutionary relationships among species.

Hence, speciation is a crucial process in the history of life on Earth, driving the generation of new species and the diversity of life forms. It occurs through the isolation, genetic divergence, and reproductive barriers that develop between populations over time. Speciation provides the raw material for evolutionary change and is essential for our understanding of the natural world.

1.1.5. Fossil Record

The fossil record is a rich and invaluable source of evidence for understanding the history of life on Earth. It comprises the preserved remains, traces, or imprints of organisms that lived in the distant past. Fossils provide critical insights into the diversity of ancient life, the evolutionary processes that shaped species, and the geological history of our planet (Figure 1.6).

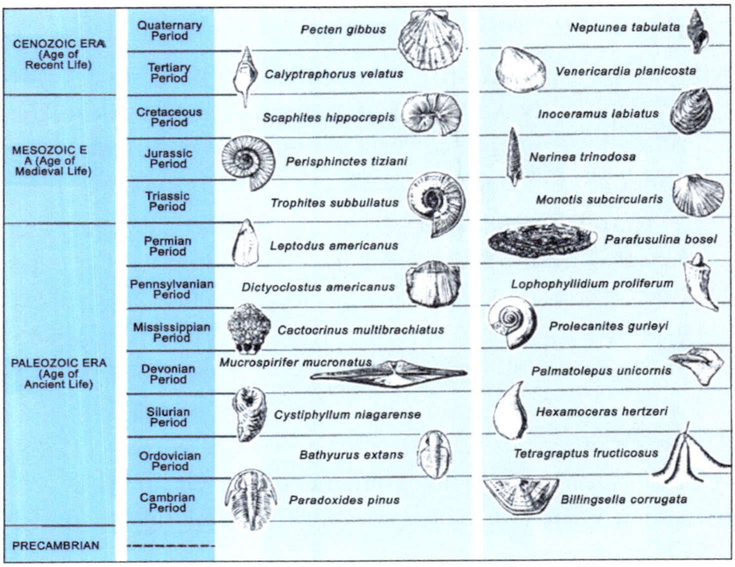

Figure 1.6. Fossil record.

Source: *By United States Geological Survey – http://pubs.usgs.gov/gip/geotime/fossils.html, Public Domain, https://commons.wikimedia.org/w/index.php?curid=3092986; Creative Commons License.*

1.1.5.1. Key Aspects of the Fossil Record

1. **Formation of Fossils:**

i. **Mineralization:** The most common way fossils form is through mineralization. When an organism dies, its remains may become buried in sediment (such as mud or sand) or submerged in water. Over time, minerals from the surrounding environment replace the organic materials in the remains, effectively turning them into stone. This process preserves the structure and sometimes even the original tissues of the organism.

ii. **Imprints and Molds:** Fossils can also form imprints or molds. For instance, when an organism's body decays or is removed from sediment, it can leave behind an impression or mold in the surrounding rock. These imprints can later be filled with minerals, creating a replica of the original organism.

iii. **Amber and Tar:** In some cases, organisms become trapped in substances like amber (fossilized tree resin) or tar pits. These substances can preserve the organism intact, including soft tissues, feathers, or fur, in remarkable detail.

iv. **Frozen Fossils:** In rare instances, fossils are preserved through freezing. Organisms that become trapped in ice can remain remarkably well-preserved over thousands or even millions of years.

2. **Types of Fossils:**

i. **Body Fossils:** These fossils preserve parts of the actual organisms, such as bones, shells, teeth, and soft tissues.

ii. **Trace Fossils:** These are indirect evidence of ancient life. They include footprints, burrows, nests, and other marks left by organisms. These provide insight into behavior and ecological interactions.

iii. **Petrified Wood:** It is a specific type of fossil where the organic material in wood is replaced by minerals, often preserving the original structure and appearance of the wood.

iv. **Microfossils:** These are extremely small fossils, including things like tiny shell fragments, pollen grains, and microorganisms. They are critical for understanding ancient ecosystems and environmental conditions.

3. **Uses and Significance of the Fossil Record:**

i. **Dating Earth's History:** Fossils play a crucial role in relative dating, helping geologists establish the sequence of rock layers (stratigraphy) and estimate the ages of rocks and fossils. This, in turn, contributes to understanding the Earth's geological history.

ii. **Understanding Evolution:** Fossils provide direct evidence of extinct species and transitional forms, helping to reconstruct the evolutionary history of organisms. They show how species have changed over time and how new species have emerged.

iii. **Paleoecology:** Fossils reveal information about ancient ecosystems, including the types of organisms that lived together, their ecological roles, and their interactions.

iv. **Biogeography:** Fossils provide insights into the distribution of organisms in the past and how continents and environments have changed over time.

v. **Human Evolution:** Fossils have been instrumental in tracing the evolution of humans and our hominin ancestors, helping to piece together our evolutionary history.

4. **Limitations of the Fossil Record:**

i. **Incomplete Preservation:** Fossilization is a rare and selective process. Most organisms do not become fossils because their remains decay or are destroyed before fossilization can occur.

ii. **Bias Toward Hard Tissues:** Fossils are more commonly found for organisms with hard parts, like bones or shells, making it more challenging to study soft-bodied organisms.

iii. **Gaps in the Record:** There are gaps in the fossil record, often due to factors like erosion, tectonic activity, or the lack of suitable rock formations for fossil preservation.

iv. **Temporal Resolution:** Fossils provide a snapshot of life at specific points in time, and the time intervals between fossil layers can vary widely. This limits our ability to track short-term evolutionary changes.

Therefore, the fossil record is a treasure trove of information about the history of life on Earth. It serves as a window into ancient ecosystems, the evolution of species, and the geological processes that have shaped our planet. While it has its limitations, the study of fossils continues to yield new discoveries and insights, deepening our understanding of the natural world.

1.1.6. Molecular Evidence

Molecular evidence refers to the information about the genetic makeup and molecular characteristics of organisms that is used to study their evolutionary relationships, genetic diversity, and relatedness to common ancestors.

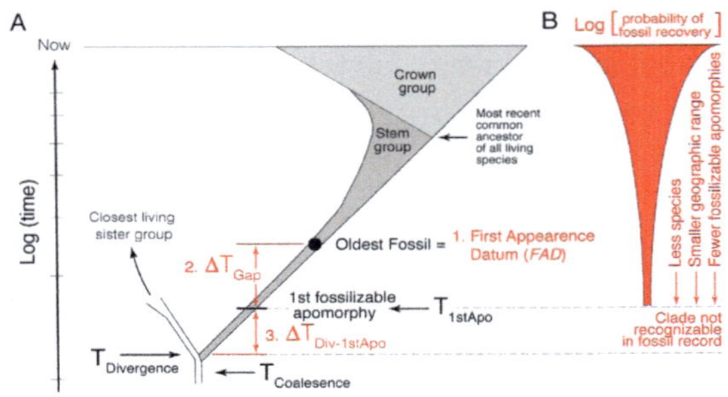

Figure 1.7. Molecular evidence.

Source: *Marshall, C. R., (2019). Using the fossil record to evaluate time-tree timescales. Front. Genet., 10, 1049. doi: 10.3389/fgene.2019.01049; Creative Commons License.*

It plays a crucial role in modern biology and has provided compelling insights into the evolutionary history of life on Earth (Figure 1.7).

1.1.6.1. Key Aspects of Molecular Evidence

1. **Genetic Material:** Molecular evidence primarily involves the study of genetic material, including DNA (deoxyribonucleic acid) and RNA (ribonucleic acid). These molecules carry the genetic information that determines the traits and characteristics of living organisms.
2. **Comparative Genomics:** DNA Sequencing: Advances in DNA sequencing technology have enabled scientists to determine the complete genetic code (genome) of numerous species. This wealth of genetic data allows for detailed comparisons of DNA sequences between different organisms.
3. **Protein Sequences:** Molecular evidence also involves the comparison of protein sequences, as proteins are the functional products of genes. The sequences of amino acids in proteins can be used to infer evolutionary relationships.

1.1.6.2. Phylogenetics

1. **Phylogenetics:** It is the field of biology that uses molecular data to reconstruct the evolutionary relationships among species and construct phylogenetic trees (family trees) that depict common ancestry.
2. **Phylogenetic Trees:** These trees illustrate the branching patterns of species, with the tips representing modern species and the nodes representing common ancestors. The length and topology of branches are determined based on the degree of genetic similarity or difference.

1.1.6.3. Molecular Clocks

Molecular evidence allows scientists to estimate the timing of evolutionary events through the use of molecular clocks. These clocks are based on the idea that genetic mutations accumulate at a relatively constant rate over time. By measuring the number of genetic differences between species, scientists can estimate how long ago their common ancestors lived.

1.1.6.4. Comparative Genomic Analyses

Molecular evidence provides insights into specific genes, gene families, and functional elements shared among species. Comparative genomics helps identify conserved genes, regulatory elements, and pathways that have been preserved throughout evolutionary history.

1.1.6.5. Significance of Molecular Evidence

1. **Reconstructing Phylogenies:** Molecular evidence has revolutionized our ability to construct accurate phylogenetic trees, providing insights into the evolutionary relationships among species. These trees are used to infer the evolutionary history of organisms, including the timing of branching events and common ancestors.
2. **Resolving Taxonomic Uncertainty:** Molecular data can help resolve taxonomic uncertainties, particularly in cases where morphology alone may not be sufficient to classify or identify species. DNA sequencing can reveal hidden genetic diversity or clarify the relationships among cryptic species.
3. **Understanding Evolutionary Mechanisms:** Molecular evidence allows researchers to study the genetic changes that underlie evolutionary processes, such as adaptation, speciation, and genetic drift. It provides a molecular basis for investigating the mechanisms of evolution.
4. **Conservation Biology:** Molecular evidence is essential in conservation biology for assessing genetic diversity within endangered populations and determining the most appropriate conservation strategies to preserve genetic variation.
5. **Biomedical Applications:** Molecular evidence has practical applications in fields like medicine and biotechnology. It helps researchers understand genetic disorders, develop treatments, and study the evolution of disease-causing organisms.

1.1.6.6. Limitations of Molecular Evidence

1. **Incomplete Data:** Molecular evidence relies on available genetic data, and some species may have limited genomic information, making it challenging to include them in phylogenetic analyses.

2. **Convergent Evolution:** It can lead to similar genetic sequences or characteristics in unrelated species due to adaptation to similar environments, potentially confounding evolutionary analyses.
3. **Horizontal Gene Transfer:** In some cases, genes can be transferred between species, complicating the interpretation of evolutionary relationships based solely on vertical inheritance.
4. **Mutation Rates:** The assumption of a constant mutation rate in molecular clocks may not always hold true, leading to inaccuracies in estimating divergence times.

In summary, molecular evidence is a powerful tool in the study of evolution and genetics. It has provided crucial insights into the relatedness of species, the timing of evolutionary events, and the genetic mechanisms underlying adaptation and diversification. It continues to play a pivotal role in advancing our understanding of the natural world and its evolutionary history.

1.1.7. The Broader Implications of Evolution (Figure 1.8)

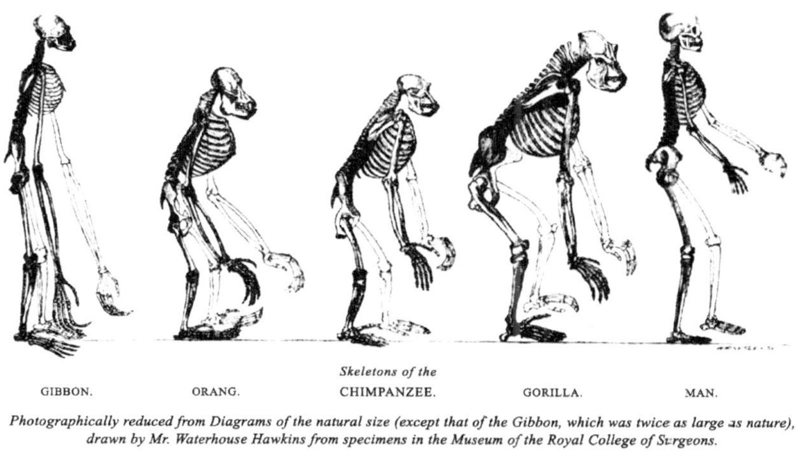

Figure 1.8. Broader implications of evolution.

Source: By Benjamin Waterhouse Hawkins (1807–94) – Frontispiece to Huxley's Evidence as to Man's Place in Nature (1863), Public Domain, https://commons.wikimedia.org/w/index.php?curid=117533; Creative Commons License.

1.1.7.1. Understanding the Diversity of Life

Evolution provides the framework for understanding the vast diversity of life on Earth. It explains how species have adapted to different environments and how they are interconnected through common ancestry. This knowledge fosters a deeper appreciation for the rich tapestry of life forms on our planet.

1. **Medical Advances:** Evolutionary biology has important implications for medicine. It helps us understand the origins of diseases, antibiotic resistance in bacteria, and the evolution of pathogens. This knowledge informs strategies for disease prevention, vaccine development, and the treatment of genetic disorders.
2. **Agriculture and Food Security:** Evolutionary principles are applied in agriculture to breed crops and livestock for desirable traits, such as higher yields, disease resistance, and improved nutritional content. This contributes to global food security and sustainable agriculture.
3. **Conservation and Biodiversity:** Evolutionary biology plays a crucial role in conservation efforts. Understanding the evolutionary history of species helps conservationists prioritize endangered species for protection, manage genetic diversity, and restore ecosystems. It also informs strategies for combating invasive species.

1.1.7.2. Biotechnology and Genetic Engineering

Biotechnology leverages our understanding of genetics and evolution to develop new technologies. Genetic engineering, for instance, allows scientists to modify organisms for various purposes, including producing pharmaceuticals, enhancing crop traits, and developing biofuels.

1.1.7.3. Human Health and Behavior

Evolutionary psychology explores how human behaviors and traits have evolved over time. This field offers insights into topics such as mate selection, social cooperation, and decision-making, shedding light on the deep-seated origins of human behavior.

1.1.7.4. Environmental Conservation

Evolutionary thinking informs conservation biology by emphasizing the importance of preserving genetic diversity within populations. Healthy genetic diversity increases the chances that species can adapt to changing environments and survive long-term challenges.

1.1.7.5. Education and Science Literacy

Evolution is a foundational concept in biology education. Teaching evolution enhances scientific literacy, critical thinking, and an understanding of the scientific method. It also fosters an appreciation for the natural world and the importance of evidence-based reasoning.

1.1.7.6. Ethical and Philosophical Considerations

Evolutionary biology has raised ethical and philosophical questions about our place in the natural world. It challenges traditional views of human uniqueness and our relationship to other species. These discussions have implications for ethics, morality, and our responsibilities toward the environment and other species.

1.1.7.7. Cultural and Societal Impact

The acceptance of evolution has far-reaching cultural and societal implications. It can influence public policy, attitudes toward science, and the way societies approach issues like climate change, resource management, and conservation.

1.1.7.8. Interdisciplinary Research

Evolutionary biology is inherently interdisciplinary, drawing from genetics, ecology, paleontology, anthropology, and other fields. This fosters collaboration and cross-disciplinary research, leading to a more holistic understanding of life and its complexities.

1.1.7.9. Global Challenges

In the face of global challenges such as climate change and emerging diseases, evolutionary principles are crucial for understanding how organisms respond to environmental shifts and how ecosystems adapt to new conditions. This knowledge informs strategies for mitigating and adapting to these challenges.

So, the broader implications of evolution extend far beyond biology. They encompass diverse areas of science, medicine, conservation, agriculture, ethics, and culture. Evolutionary principles offer valuable insights and tools for addressing many of the pressing challenges and questions facing our world today.

1.2. THE SIGNIFICANCE OF THE EVOLUTIONARY THEORY

1. **Unifying Biological Science:** Evolution serves as the central unifying principle of biology. It provides a coherent framework that ties together diverse biological disciplines, including genetics, ecology, paleontology, and microbiology. By understanding the processes of evolution, scientists can make sense of the complexity and diversity of life.
2. **Explaining Biodiversity:** Evolutionary theory explains the origin and proliferation of biodiversity on Earth. It helps us understand how species have adapted to various environments, leading to the incredible variety of life forms we observe today.
3 **Reconstructing Life's History:** Evolutionary theory allows scientists to reconstruct the history of life on Earth. By examining the fossil record and using molecular evidence, researchers can trace the evolutionary relationships among species and understand the temporal sequence of events in the history of life.
4 **Practical Applications in Medicine:** Evolutionary biology has practical applications in medicine, such as understanding the evolution of pathogens, antibiotic resistance, and the genetic basis of diseases. This knowledge informs disease prevention, vaccine development, and personalized medicine.
5. **Agricultural Advancements:** Evolutionary principles are applied in agriculture to breed crops and livestock for desired traits, such as higher yields and disease resistance. This contributes to global food security and sustainable farming practices.
6. **Conservation and Ecosystem Management:** Evolutionary insights are crucial in conservation biology. Understanding the evolutionary history of species helps prioritize conservation efforts, manage genetic diversity within populations, and restore ecosystems.
7. **Biotechnology and Genetic Engineering:** Evolutionary knowledge informs biotechnology and genetic engineering, enabling the modification of organisms for various purposes,

including producing pharmaceuticals, enhancing crop traits, and developing biofuels.

8. **Human Health and Behavior:** Evolutionary psychology explores how human behaviors and traits have evolved over time. This field offers insights into topics such as mate selection, social cooperation, and decision-making, contributing to our understanding of human behavior.
9. **Environmental Conservation:** Evolutionary thinking guides environmental conservation strategies by emphasizing the importance of preserving genetic diversity within populations. Healthy genetic diversity increases the adaptability of species to changing environments.
10. **Science Literacy and Education:** Evolution is a cornerstone of biology education. Teaching evolution promotes scientific literacy, critical thinking, and an understanding of the scientific method. It fosters an appreciation for the natural world and evidence-based reasoning.
11. **Ethical and Philosophical Implications:** Evolutionary biology raises ethical and philosophical questions about our place in the natural world. It challenges traditional views of human uniqueness and our relationship with other species, influencing discussions about ethics, morality, and environmental responsibility.
12. **Cultural and Societal Impact:** The acceptance of evolution has profound cultural and societal implications. It can influence public policy, attitudes toward science, and the way societies approach issues like climate change, resource management, and conservation.
13. **Interdisciplinary Research:** Evolutionary biology encourages interdisciplinary research, fostering collaboration across fields such as genetics, ecology, anthropology, and more. This interdisciplinary approach leads to a more holistic understanding of life and its complexities.
14. **Addressing Global Challenges:** Evolutionary principles are essential for understanding how organisms respond to global challenges like climate change, emerging diseases, and habitat loss. This knowledge informs strategies for mitigating and adapting to these challenges.

Thus, evolutionary theory is not merely a biological concept it is a foundational principle that has far-reaching implications for science, medicine, agriculture, conservation, ethics, culture, and society. Its significance extends across multiple domains and contributes to our understanding of the natural world and our role within it (Figure 1.9).

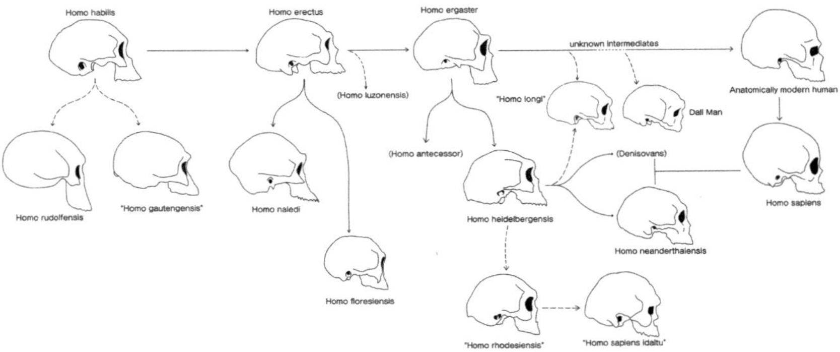

Figure 1.9. Significance of the evolutionary theory.

Source: *By SimplisticReps – Own work, CC BY-SA 4.0, https://commons.wikimedia.org/w/index.php?curid=112814051; Creative Commons License.*

REFERENCES

1. Appleman, P., (1970). *Darwin*. New York: Norton.
2. Bowler, P. J., (1975). The changing meaning of 'evolution.' *Journal of the History of Ideas, 36*, 95–114.
3. Browne, J., (1995–2002). *Charles Darwin: A Biography* (Vol. 2). New York: Knopf.
4. Burkhardt, R. W. Jr., (1977). *The Spirit of System: Lamarck and Evolutionary Biology*. Cambridge, Mass: Harvard University Press.
5. Charles, D. *On the Origin of Species.*
6. Darwin, C., (1859). *On the Origin of the Species by Means of Natural Selection: The Preservation of Favored Races in the Struggle for Life*. London: J. Murray.
7. David, Q. *The Tangled Tree: A Radical New History of Life*.
8. Desmond, A. J., & James, R. M., (1991). *Darwin*. London: Michael Joseph.
9. Desmond, A. J., (1989). *The Politics of Evolution: Morphology, Medicine, and Reform in Radical London*. Chicago: University of Chicago Press.
10. Dobzhansky, T. G., (1937). *Genetics and the Origin of Species*. New York: Columbia University Press.
11. *Evolution: The History of an Idea*, (2003). 3rd edn. Berkeley: University of California Press.
12. Fichman, M., (2004). *An Elusive Victorian: The Evolution of Alfred Russel Wallace*. Chicago: University of Chicago Press.
13. Gould, S. J., (2002). *The Structure of Evolutionary Theory*. Cambridge, Mass: Belknap Press.
14. Greene, J. C., (1996). *The Death of Adam: Evolution and Its Impact on Western Thought* (rev. edn.). Ames: Iowa State University Press.
15. Huxley, J., (1963). *Evolution: The Modern Synthesis*. London: Allen and Unwin.
16. Jerry, A. C. *Why Evolution is True*.
17. Jonathan, B. L. *Improbable Destinies: Fate, Chance, and the Future of Evolution*.
18. Jonathan, W. *The Beak of the Finch: A Story of Evolution in Our Time*.

19. Kohn, D., (1985). *The Darwinian Heritage*. Princeton, N.J.: Princeton University Press.
20. Mayr, E., & William, B. P., (1980 & 2000). *The Evolutionary Synthesis: Perspectives on the Unification of Biology*. Cambridge, Mass: Harvard University Press.
21. Mayr, E., (1982). *The Growth of Biological Thought: Diversity, Evolution, and Inheritance*. Cambridge, Mass: Belknap Press.
22. *Monad to Man: The Concept of Progress in Evolutionary Biology*, (1996). Cambridge, Mass: Harvard University Press.
23. Neil, S., (2009). *Your Inner Fish: A Journey into the 3.5-Billion-Year History of the Human Body*. Vintage Books; 0307277453, 9780307277459.
24. Raby, P., (2001). *Alfred Russel Wallace: A Life*. Princeton, N.J.: Princeton University Press.
25. Rebecca, S. *Darwin's Ghosts: The Secret History of Evolution*.
26. Richard, D. *The Blind Watchmaker*.
27. Richard, D. *The Extended Phenotype*.
28. Richard, D. *The Greatest Show on Earth*.
29. Richard, D. *The Selfish Gene*.
30. Richards, R. J., (1992). Evolution. In: Evelyn, F. K., & Elizabeth, L., (eds.), *Keywords in Evolutionary Biology* (pp. 95–105). Cambridge, Mass: Harvard University Press.
31. Ruse, M., (1979). *The Darwinian Revolution: Science Red in Tooth and Claw* (2nd edn.). Chicago: University of Chicago Press.
32. Sean, B. C. *Endless Forms Most Beautiful: The New Science of Evo Devo*.
33. Sean, B. C. *The Serengeti Rules: The Quest to Discover How Life Works and Why It Matters*.
34. Secord, J. A., (2000). *Victorian Sensation: The Extraordinary Publication, Reception, and Secret Authorship of Vestiges of the Natural History of Creation*. Chicago: University of Chicago Press.
35. Shermer, M., (2002). *In Darwin's Shadow: The Life and Science of Alfred Russel Wallace*. A biographical study on the psychology of history. Oxford and New York: Oxford University Press.
36. Smocovitis, V. B., (1996). *Unifying Biology: The Evolutionary Synthesis and Evolutionary Biology*. Princeton, N.J.: Princeton University Press.

37. *The Eclipse of Darwinism: Anti-Darwinian Evolution Theories in the Decades around 1900*, (1983). Baltimore: Johns Hopkins University Press.

38. *The Meaning of Evolution: The Morphological Construction and Ideological Reconstruction of Darwin's Theory*, (1992). Chicago: University of Chicago Press.

39. The Non-Darwinian Revolution, (1988). *Reinterpreting a Historical Myth*. Baltimore: Johns Hopkins University Press.

40. *The Romantic Conception of Life: Science and Philosophy in the Age of Goethe*, (2002). Chicago: University of Chicago Press.

41. Wallace, A. R., (1881). *Island Life*. Amherst, N.Y.: Prometheus Books–1998.

42. *What Evolution Is*, (2001). New York: Basic Books.

43. Young, R. M., (1985). *Darwin's Metaphor: Nature's Place in Victorian Culture*. Cambridge, U.K., and New York: Cambridge University Press.

44. Yuval, N. H. *Sapiens: A Brief History of Humankind.*

CHAPTER 2

Early Ideas and Precursors to Evolution

CONTENTS

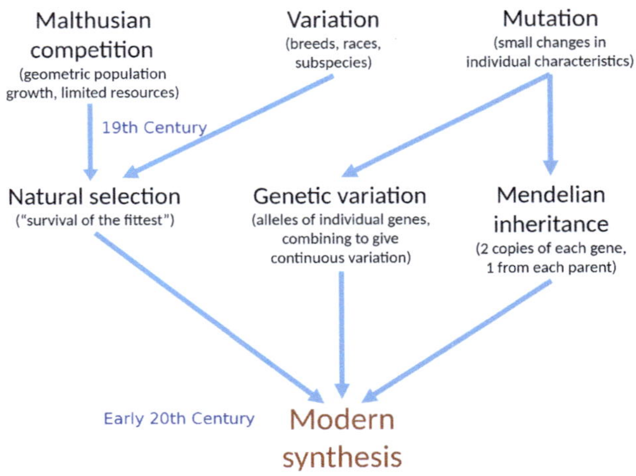

Figure 2.1. History of evolution thought.

Source: *By Ian Alexander – Own work, CC BY-SA 4.0, https://commons.wikimedia.org/w/index.php?curid=62385744; Creative Commons License.*

2.1. PRE-DARWINIAN THEORIES

Before Charles Darwin's groundbreaking work, several theories and ideas circulated about the origins and development of life. These early theories, though often incomplete or speculative, laid the foundation for the concept of evolution. Here's an overview of some key pre-Darwinian theories:

1. **Ancient Philosophical Ideas:**

i. **Anaximander (610–546 BCE):** Anaximander, a Greek philosopher, proposed the idea that life originated in water and that simpler forms of life gradually transformed into more complex forms.

ii. **Empedocles (c. 490–430 BCE):** Empedocles, another Greek philosopher, suggested that various body parts or organisms could combine in different ways, leading to the formation of new life forms.

2. **Great Chain of Being:** It was a hierarchical structure prevalent in medieval and Renaissance thought. It depicted a linear order of life forms from the simplest to the most complex, with God at the top. While it didn't propose evolutionary change, it influenced early biological thinking.

3. **Transformational Ideas:**

i. **Jean-Baptiste Lamarck (1744–1829):** Lamarck proposed the theory of inheritance of acquired characteristics. He suggested that organisms could change during their lifetimes and pass on these changes to their offspring, driving evolutionary adaptation.

ii. **Erasmus Darwin (1731–1802):** Charles Darwin's grandfather, Erasmus, suggested that all forms of life shared a common ancestry and that life had evolved over time, although he didn't provide a comprehensive mechanism for this.

4. **Catastrophism and Uniformitarianism:**

i. **Georges Cuvier (1769–1832):** Cuvier was a proponent of catastrophism, suggesting that Earth's geological features were shaped by sudden, short-lived events. However, he acknowledged the extinction of species, laying the groundwork for the understanding that life on Earth has changed over time.

ii. **James Hutton (1726–1797) and Charles Lyell (1797–1875):** Hutton and Lyell proposed the principle of uniformitarianism, suggesting that Earth's geological processes operated at a consistent rate over time. This idea helped establish the concept of deep time, allowing for the slow, gradual changes necessary for evolution.

5. **Transmutation of Species:**

i. **Etienne Geoffroy Saint-Hilaire (1772–1844):** Geoffroy proposed the theory of unity of composition, suggesting that animals shared common structural elements, indicating a common ancestry. He challenged the fixity of species.

ii. **Alfred Russel Wallace (1823–1913):** Wallace independently formulated a theory of natural selection and sent his ideas to Darwin, leading to the joint presentation of the theory of evolution by natural selection.

These pre-Darwinian ideas, though disparate and often lacking a unifying framework, provided the intellectual backdrop against which Darwin and Wallace formulated the theory of evolution by natural selection, revolutionizing the understanding of life's diversity and its origins.

2.1.1. Renaissance and Enlightenment

During the Renaissance and Enlightenment eras, scientific thinking and inquiry began to challenge prevailing religious dogmas. Scholars and naturalists of this period laid the groundwork for evolutionary thought through observations, theories, and inquiries, paving the way for the eventual development of Darwin's theory of evolution by natural selection.

1. **Leonardo da Vinci (1452–1519) (Figure 2.2):**

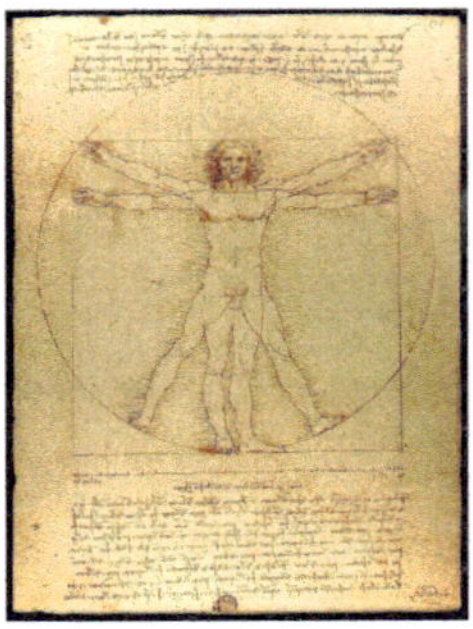

Figure 2.2. Leonardo's extensive studies of anatomy and observations of living organisms.

Source: *By Leonardo da Vinci – Leonardo Da Vinci – Photo from www.lucnix.be. 2007–09–08 (photograph). Photography: This image is the work of Luc Viatour Please credit it with: Luc Viatour/https://Lucnix.be in the immediate vicinity of the image. https://commons.wikimedia.org/w/index.php?curid=2738140; Creative Commons License.*

2. **Carolus Linnaeus (1707–1778) (Figure 2.3):**

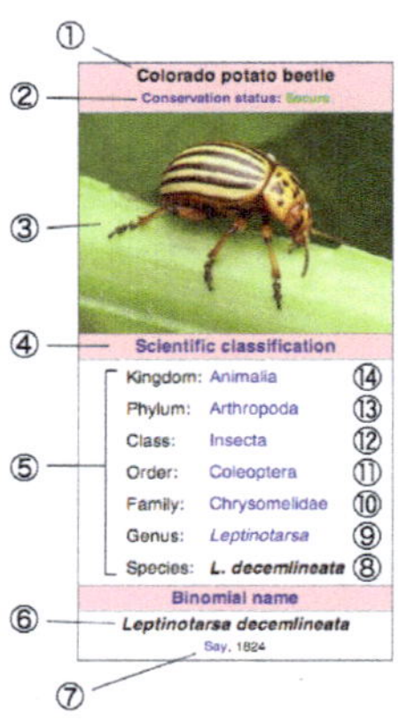

Figure 2.3. Carolus Linnaeus's binomial nomenclature system.

Source: CC BY-SA 3.0, https://commons.wikimedia.org/w/index.php?curid=484234; Creative Commons License.

Linnaeus, often known as the father of modern taxonomy, developed the binomial nomenclature system for classifying and naming organisms. While not an evolutionary theorist, his work provided a foundation for organizing the diversity of life, a crucial precursor to evolutionary studies.

3. **Georges-Louis Leclerc, Comte de Buffon (1707–1788) (Figure 2.4):**

Figure 2.4. Georges-Louis Leclerc, Comte de Buffon's burning mirror.

Source: *By User: Ruhrfisch – Photographed it myself, CC BY-SA 3.0, https://commons.wikimedia.org/w/index.php?curid=3266063; Creative Commons License.*

Buffon's monumental work, "Histoire Naturelle," discussed the idea that species could change over time due to environmental influences. He proposed that species might transform gradually, challenging the concept of fixed and unchanging species.

4. **Jean-Baptiste Lamarck (1744–1829) (Figure 2.5):**

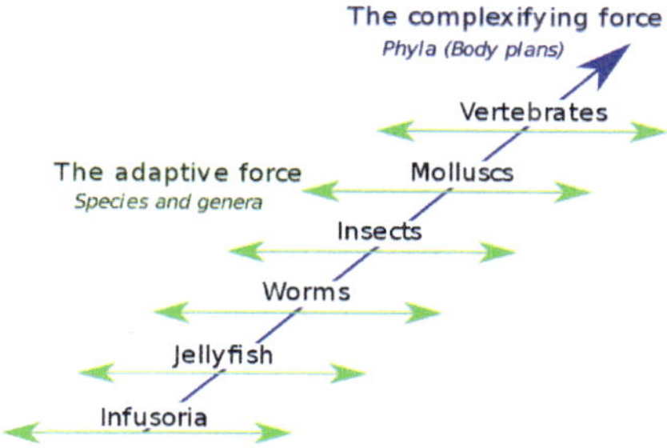

Figure 2.5. Jean-Baptiste Lamarck's evolution theory.

Source: By Ian Alexander – Own work, CC BY-SA 4.0, https://commons.wikimedia.org/w/index.php?curid=67182770; Creative Commons License.

Lamarck proposed an early theory of evolution in which he suggested that organisms could change over time through the inheritance of acquired traits. He believed that characteristics acquired during an organism's lifetime could be passed on to the next generation, influencing evolutionary change.

5. **Erasmus Darwin (1731–1802) (Figure 2.6):**

Figure 2.6. Erasmus Darwin romantic natural history.

Source: *By Joseph Wright of Derby – Unknown source, Public Domain, https://commons.wikimedia.org/w/index.php?curid=515673; Creative Commons License.*

Charles Darwin's grandfather, Erasmus, was a respected physician, naturalist, and poet. In his work, "Zoonomia," he speculated about the transmutation of species, suggesting that all forms of life shared a common ancestry. Although speculative, his ideas were part of the intellectual atmosphere of the time.

6. **Pierre-Louis Maupertuis (1698–1759) (Figure 2.7):**

Figure 2.7. Pierre-Louis Maupertuis.

Source: *Public Domain, https://commons.wikimedia.org/w/index.php?curid=21900; Creative Commons License.*

Maupertuis proposed the principle of least action, which later influenced the concept of natural selection. He suggested that the forces of nature tend to act in the most economical way, a notion that resonates with the efficiency of evolutionary adaptations.

7. **Charles Bonnet (1720–1793) (Figure 2.8):**

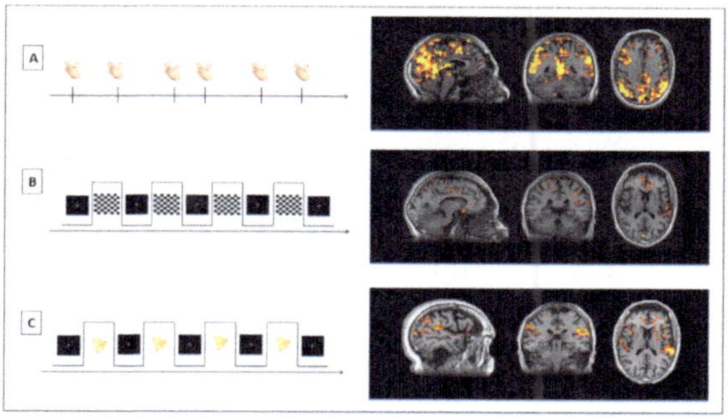

Figure 2.8. Charles Bonnet syndrome.

Source: Vacchiano, V., Tonon, C., Mitolo, M., et al., (2019). Functional MRI study in a case of Charles Bonnet syndrome related to LHON. BMC Neurol., 19, 350. https://doi.org/10.1186/s12883-019-1579-9; Creative Commons License.

Bonnet's work on reproduction and development in insects led him to propose the idea of preformationism, suggesting that organisms contained miniature versions of themselves in their reproductive cells. While his ideas were not evolutionary in the modern sense, they contributed to the understanding of biological development.

8. **Georges Cuvier (1769–1832) (Figure 2.9):**

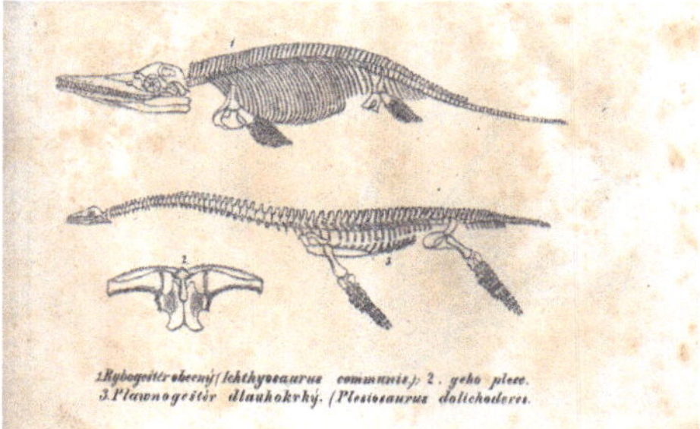

Figure 2.9. Georges Cuvier on fossils and extinctions.

Source: By HTO – my archive, Public Domain, https://commons.wikimedia.org/w/index.php?curid=5932161; Creative Commons License.

Cuvier, a prominent naturalist, and paleontologist, argued for the reality of extinction. His studies of fossils and comparative anatomy helped establish the idea that species could disappear from the Earth, setting the stage for the concept of a changing, dynamic Earth with evolving life forms. During the Renaissance and Enlightenment periods, these thinkers laid the foundation for evolutionary thought by challenging prevailing notions of fixity and advocating for the study of nature through observation and systematic classification. While their ideas were often speculative and lacked a coherent theory of evolution, they contributed to the intellectual climate that eventually led to the development of Darwin's theory of evolution by natural selection.

2.1.2. Transmutation of Species

Before the publication of Charles Darwin's "On the Origin of Species" in 1859, the idea of transmutation of species, which suggests that species change over time and transform into new species, was a topic of debate among naturalists. Here are some key figures and their contributions to the concept of transmutation of species:

1. **Jean-Baptiste Lamarck (1744–1829):** Lamarck proposed an early theory of evolution that emphasized the inheritance of acquired characteristics. He believed that organisms could change during their lifetimes in response to their environment, and these acquired traits could be passed on to offspring. While Lamarck's specific mechanism of evolution was later discredited, his ideas contributed to the broader concept of species change over time (Figure 2.10).

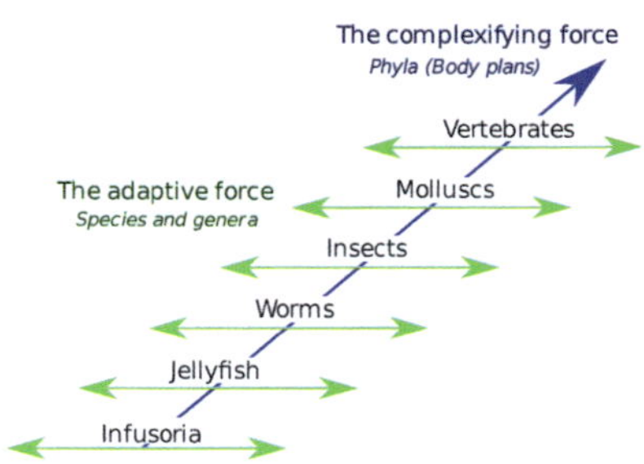

Figure 2.10. Jean-Baptiste Lamarck.

Source: scihi.org; Creative Commons License.

2 **Erasmus Darwin (1731–1802):** Charles Darwin's grandfather, Erasmus, suggested the idea of common ancestry and the transmutation of species in his work "Zoonomia." He proposed that all forms of life shared a common origin, and that species were not fixed but could change over time.

3. **Geoffroy Saint-Hilaire (1772–1844):** Etienne Geoffroy Saint-Hilaire, a French naturalist, proposed the theory of unity of composition, suggesting that animals shared common structural elements, indicating a common ancestry. He challenged the fixity of species and argued for a more dynamic view of life's transformations (Figure 2.11).

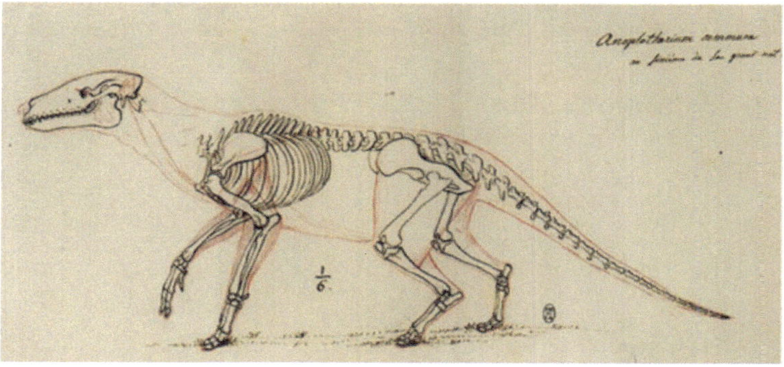

Figure 2.11. Geoffroy Saint-Hilaire.

Source: *By Georges Cuvier – Reviewing the iconography and the central role of 'paleoart': four centuries of geo-paleontological art: Figure 22 (Listed as Public Domain/CC0 due to being produced by Georges Cuvier around 1812). Public Domain, https://commons.wikimedia.org/w/index.php?curid=135370940; Creative Commons License.*

4. **Alfred Russel Wallace (1823–1913):** While not a proponent of transmutation in the same sense as Lamarck or Darwin, Wallace independently formulated a theory of natural selection similar to Darwin's. In 1858, he sent his ideas to Darwin, leading to the joint presentation of the theory of evolution by natural selection (Figure 2.12).

Figure 2.12. Alfred Russel Wallace.

Source: *By Alfred Russel Wallace (1823–1913) – https://pixels.com/featured/sugar-palm-natural-history-museum-london.html, Public Domain, https://commons.wikimedia.org/w/index.php?curid=84848945; Creative Commons License.*

5. **Robert Chambers (1802–1871):** Chambers anonymously published the book "Vestiges of the Natural History of Creation" in 1844, which proposed a speculative evolutionary theory. The book suggested a progressive development of life forms, culminating in the emergence of humans. Though controversial and lacking scientific rigor, it sparked public interest in the idea of transmutation of species (Figure 2.13).

Figure 2.13. Robert Chambers.

Source: *By Unknown author – British library, Public Domain, https://commons.wikimedia.org/w/index.php?curid=10397469; Creative Commons License.*

These early evolutionary ideas, along with the emerging fields of geology and paleontology, laid the groundwork for the acceptance of Darwin's theory of evolution by natural selection. Darwin's meticulous research, extensive evidence gathering, and comprehensive argumentation in "On the Origin of Species" provided a robust scientific foundation for the concept of transmutation of species, shaping the course of modern biology.

2.1.3. David Hume

David Hume (1711–1776) was an influential Scottish philosopher, historian, economist, and essayist known for his empiricism and skepticism. His works profoundly shaped modern philosophy, especially in the areas of empiricism, skepticism, and the philosophy of science. Here are some key aspects of his contributions (Figure 2.14):

Figure 2.14. David Hume.

Source: By Allan Ramsay – https://www.nationalgalleries.org/art-and-artists/60610/david-hume-1711-1776-historian-and-philosopher-1754, Public Domain, https://commons.wikimedia.org/w/index.php?curid=1367760; Creative Commons License.

1. **Empiricism and Empirical Philosophy:** Hume was a staunch empiricist, emphasizing the importance of sensory experience as the foundation of human knowledge. He argued that all meaningful ideas and concepts are derived from experience, challenging notions of innate ideas.

2. **Problem of Induction:** One of Hume's most significant contributions to philosophy was his problem of induction. He argued that the justification for inductive reasoning, crucial in scientific inference, cannot be rationally justified. This philosophical problem highlighted the challenges in drawing general conclusions from specific observations.
3. **Causality and Skepticism:** Hume's philosophical investigations led him to question the concept of causality. He argued that our belief in cause-and-effect relationships is not derived from reason but rather from habit or custom. This skepticism about causality had a profound impact on subsequent philosophy and scientific thought.
4. **Moral Philosophy:** In his moral philosophy, particularly in his work "A Treatise of Human Nature," Hume argued for a sentiment-based approach to ethics. He posited that moral judgments are derived from sentiments, such as approval or disapproval, rather than reason alone. This approach laid the groundwork for modern emotivism in ethics.
5. **Philosophy of Religion:** Hume was critical of religious beliefs and miracles. In his essay "Of Miracles," he argued that extraordinary claims, such as miracles, should be met with extraordinary evidence. He challenged the credibility of miracle reports, questioning their rational acceptance.
6. **History and Politics:** Besides his philosophical works, Hume also made significant contributions to history and political theory. His "History of England" became a standard reference and influenced subsequent historical writing. In political theory, he advocated for a balanced constitution and the importance of stable government.

Hume's empiricism, skepticism, and emphasis on empirical observation profoundly influenced the development of philosophy and science. His ideas continue to be relevant in contemporary discussions on knowledge, causality, ethics, and the philosophy of science.

2.1.4. Dialogues Concerning Natural Religion

"Dialogues Concerning Natural Religion" is a philosophical work by the Scottish philosopher David Hume. In this posthumously published text, Hume explores complex philosophical questions about the existence and nature of God, the problem of evil, and the limits of human knowledge. The dialogue format of the book presents these ideas through the voices of three characters: Cleanthes, Philo, and Demea.

1. **Cleanthes:** It represents the viewpoint of a natural theologian. He argues for the existence of God based on the design and order found in the natural world. Cleanthes uses the argument from design, asserting that the complexity and beauty of the universe imply the existence of an intelligent creator.
2. **Philo:** A skeptic and philosophical critic, presents a series of objections against the arguments for God's existence. He questions the validity of analogies used in the argument from design, highlighting the limitations of human understanding when trying to infer the nature of the divine.
3. **Demea:** It represents a more orthodox and traditional religious perspective. He advocates for a rational approach to understanding God, emphasizing the limitations of human reason and the need for faith.

Throughout the dialogues, Hume does not explicitly endorse any particular viewpoint but instead uses the characters to present various arguments and counterarguments. The work raises profound questions about the nature of religious belief, the validity of religious arguments, and the extent of human knowledge in matters of theology.

2.1.4.1. Key Themes and Ideas

1. **Problem of Evil:** The dialogues address the classic problem of evil, questioning how the existence of a benevolent and omnipotent God can be reconciled with the existence of suffering and evil in the world.
2. **Limits of Human Reason:** Hume explores the limitations of human reason when attempting to understand the divine. The dialogues emphasize the challenges of drawing conclusions about God's nature based solely on observations of the natural world.
3. **Critique of Religious Arguments:** Hume critically examines the traditional arguments for the existence of God, particularly the argument from design. He raises questions about the validity of analogies drawn between human-made objects and the natural world.

"Dialogues Concerning Natural Religion" continues to be a significant work in the philosophy of religion, challenging readers to critically assess the foundations of religious belief and the rationality of theological arguments.

2.2. CONTRIBUTIONS OF LAMARCK

Jean-Baptiste Lamarck (1744–1829), a French naturalist, biologist, and philosopher, is often remembered as one of the pioneers of evolutionary theory. Although his specific mechanism of evolution, known as Lamarckism, has been largely superseded by Darwin's theory of natural selection, Lamarck's contributions to the understanding of species change and adaptation remain significant in the history of biology.

1. **Early Life and Scientific Background:** Born into a noble family, Lamarck showed early interest in the natural world. He studied medicine and botany, eventually joining the French army where he cultivated his passion for natural history. Lamarck's extensive travels and studies laid the foundation for his later scientific work.
2. **Species Change and Environmental Adaptation:** Lamarck proposed the idea that species evolve over time through a process of adaptation to their environment. He argued that the use or disuse of organs could influence their development in successive generations. This concept, known as the inheritance of acquired characteristics, formed the core of Lamarck's evolutionary theory. According to Lamarck, organisms could change during their lifetimes in response to environmental pressures, and these acquired traits would be passed on to their offspring.
3. **Lamarckian Inheritance:** Lamarck also proposed the concept of the inheritance of acquired traits, suggesting that modifications acquired during an organism's lifetime could be transmitted to its offspring. While this idea has been largely discredited in modern genetics, it was an early attempt to explain how traits could change over generations.
4. **Classification and Paleontology:** Lamarck made significant contributions to the classification of invertebrate animals. He classified numerous species based on their structural features, laying the groundwork for later advancements in taxonomy. Additionally, he conducted extensive studies in paleontology, recognizing the existence of extinct species and proposing the idea of the transmutation of species over geological time.

5. **Continuity of Life and Progression:** Lamarck argued for the continuity of life, suggesting that all living organisms were interconnected and had a shared ancestry. He believed that life had progressed from simple to complex forms, demonstrating a sense of evolutionary progression.
6. **Legacy and Critique:** While Lamarck's ideas were groundbreaking, they faced criticism during his lifetime and after. One of the main criticisms came from the scientific community's inability to understand the mechanisms underlying the inheritance of acquired traits. Additionally, Lamarck's theory lacked a mechanism comparable to natural selection, which Darwin later provided.

Jean-Baptiste Lamarck made significant contributions to the early understanding of evolution and adaptation. His ideas paved the way for future evolutionary thinkers, sparking important debates and discussions about the mechanisms driving species change. Despite the eventual shift towards Darwinian natural selection as the predominant explanation for evolution, Lamarck's work remains a foundational part of the historical development of evolutionary theory, highlighting the complexity and diversity of life on Earth.

2.3. THE INFLUENCE OF FOSSIL DISCOVERIES

Fossil discoveries have played a pivotal role in shaping our understanding of evolution, providing crucial evidence for the existence of ancient life forms and the processes that have led to the diversity of species on Earth. Here, we delve into the detailed analysis of how fossil discoveries have influenced evolutionary theory over the years.

1. **Confirmation of Extinction:** Fossil discoveries unequivocally confirmed the existence of organisms that no longer inhabit the Earth. The recognition of extinct species challenged the prevailing belief in the fixity of species, suggesting that life forms had changed over geological time. This realization was a fundamental departure from traditional views and laid the groundwork for the acceptance of evolutionary concepts.
2. **Transitional Fossils and Evolutionary Links:** The discovery of transitional fossils provided direct evidence of intermediate forms between different groups of organisms. Notable examples include the Archaeopteryx, a transitional form between reptiles

and birds, and Tiktaalik, a fish-tetrapod transitional fossil. These fossils offer tangible proof of evolutionary links, showcasing the gradual changes in anatomical features over time.

3. **Understanding Evolutionary Trends:** By studying the fossil record, scientists have identified evolutionary trends, demonstrating how certain traits evolved over millions of years. For instance, the evolution of horse ancestors from small, multi-toed creatures to the large, single-toed horses we recognize today is well-documented in the fossil record. These trends provide insights into the adaptive processes that have shaped species over geological epochs.
4. **Mass Extinctions and Adaptive Radiations:** Fossil evidence of mass extinctions, such as the Permian-Triassic and Cretaceous-Paleogene events, highlighted the dynamic nature of life on Earth. These extinction events paved the way for rapid adaptive radiations, where surviving species diversified to fill ecological niches left vacant by extinct organisms. Fossils documenting these events shed light on the mechanisms of recovery and evolution following mass extinctions.
5. **Molecular Paleontology and Ancient DNA:** Recent advancements in molecular paleontology have allowed scientists to extract and analyze ancient DNA from fossils. This revolutionary technique has provided insights into the genetic relationships between extinct species and their modern counterparts. Studying ancient DNA has revealed evolutionary changes at the genetic level, complementing the morphological information obtained from traditional fossil analysis.
6. **Challenges and Gaps in the Fossil Record:** While fossils have immensely enriched our understanding of evolutionary processes, there are still gaps and challenges in the fossil record. The absence of certain transitional forms and the biases in fossil preservation present limitations. However, ongoing discoveries and improved methods continue to address these gaps, gradually refining our knowledge of evolutionary history.

Therefore, the influence of fossil discoveries on evolutionary theory is immeasurable. Fossils provide tangible, tangible evidence of life forms that lived millions of years ago, offering a window into the past and supporting the fundamental principles of evolutionary biology. As technology advances and more fossils are unearthed, our understanding of the intricate tapestry of life's evolution will undoubtedly continue to deepen, further illuminating the processes that have shaped the vast diversity of living organisms on Earth.

REFERENCES

1. Alpheus, S. P. *Lamarck, the Founder of Evolution: His Life and Work.*
2. Aristotle. *The History of Animals.*
3. Desmond, K.-H. *Erasmus Darwin: A Life of Unequalled Achievement.*
4. Jean, G. The Transformist Illusion.
5. Jonathan, W. The Beak of the Finch: A Story of Evolution in Our Time.
6. Martin, J. S. R. *Georges Cuvier, Fossil Bones, and Geological Catastrophes: New Translations & Interpretations of the Primary Texts.*
7. Richard, C. *The Species Seekers: Heroes, Fools, and the Mad Pursuit of Life on Earth.*
8. Sean, B. C. *The Making of the Fittest: DNA and the Ultimate Forensic Record of Evolution.*
9. Thomas, S. K. *The Structure of Scientific Revolutions.*

CHAPTER 3

Charles Darwin and the Origin of Species

CONTENTS

Figure 3.1. Charles Darwin.

Source: *By George Richmond – From Origins, Richard Leakey and Roger Lewin, Public Domain, https://commons.wikimedia.org/w/index.php?curid=22092879; Creative Commons License.*

3.1. CHARLES DARWIN'S LIFE AND INFLUENCES

Charles Darwin, the eminent naturalist who revolutionized our understanding of life on Earth through his theory of evolution by natural selection, was shaped by a myriad of experiences and influences throughout his life. Here, we undertake a detailed analysis of the life and influences that molded the mind of this pioneering scientist.

1. **Early Life and Education:** Born on February 12, 1809, in Shrewsbury, England, Darwin came from a family with a strong scientific background. His father, Robert Darwin, was a successful physician, and his grandfather, Erasmus Darwin, was a respected naturalist and poet. Growing up in this intellectually stimulating environment, young Charles developed an early interest in nature.
2. **The Voyage of the Beagle:** One of the most transformative experiences in Darwin's life was his five-year voyage aboard HMS Beagle (1831–1836). As the ship's naturalist, he traveled to various continents, including South America, the Galapagos Islands, and Australia. During this expedition, Darwin made numerous observations of diverse species and geological formations, which profoundly influenced his later theories. His encounter with the Galapagos finches, in particular, provided crucial insights into adaptation and variation among species (Figure 3.2).

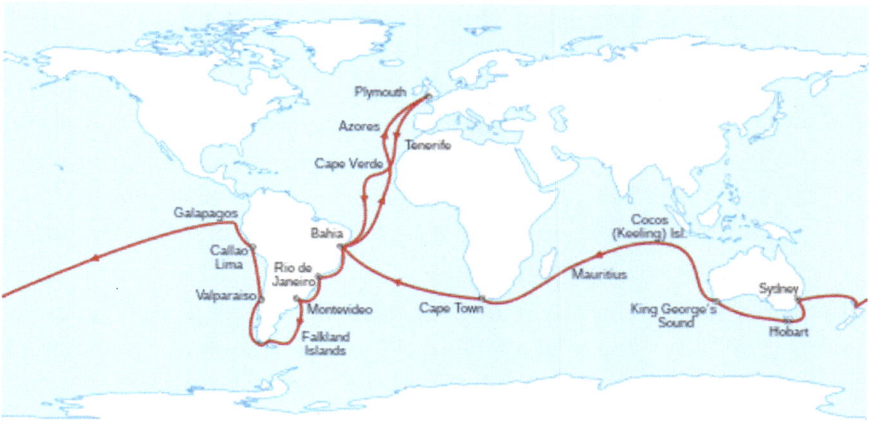

Figure 3.2. Charles Darwin's voyage of the beagle.

Source: *By © Sémhur / Wikimedia Commons, CC BY-SA 4.0, https://commons.wikimedia.org/w/index.php?curid=3003026; Creative Commons License.*

3. **Influential Scientific Figures:** Darwin was deeply influenced by the works of other naturalists and scientists of his time. He admired the writings of Alexander von Humboldt, a pioneering explorer and naturalist, and was influenced by Charles Lyell's geology, which emphasized gradual geological processes over vast spans of time. Additionally, Thomas Malthus's essay on population growth and resource scarcity inspired Darwin's thoughts on competition and natural selection.
4. **Personal Challenges and Determination:** Darwin faced significant personal challenges during his life, including chronic health issues and the loss of his daughter Annie. Despite these difficulties, his determination to pursue scientific inquiry remained unshaken. Darwin's meticulous observations, experiments, and correspondence with a network of scientists were instrumental in the development of his theory of evolution.
5. **Collaboration and Support:** Darwin benefited from collaboration and support from fellow scientists. His partnership with Joseph Hooker, the renowned botanist, provided valuable insights into the diversity of plant life. Moreover, the encouragement and critique he received from fellow naturalist Alfred Russel Wallace, who independently formulated a similar theory of evolution, prompted Darwin to publish his groundbreaking work.

6. **"On the Origin of Species" and Legacy:** In 1859, Darwin published his seminal work, "On the Origin of Species by Means of Natural Selection." This masterpiece presented a comprehensive explanation of evolution through natural selection, highlighting the role of adaptation and the struggle for existence in shaping biological diversity. The publication of this book revolutionized biology, providing a unifying framework for understanding the origins and diversity of life.

In fact, Charles Darwin's life and influences were a rich tapestry of scientific curiosity, personal challenges, and intellectual exchange. His insatiable curiosity about the natural world, coupled with his meticulous observations and critical thinking, led to the formulation of one of the most impactful scientific theories in history. Darwin's legacy endures not only in evolutionary biology but also as a testament to the power of relentless inquiry and the transformative potential of scientific discovery.

3.2. THE VOYAGE OF THE HMS BEAGLE

The HMS Beagle's voyage, spanning from 1831 to 1836, stands as one of the most influential scientific expeditions in history. Commanded by Captain Robert FitzRoy, the ship's primary mission was to survey and chart the coasts of South America, but it evolved into a groundbreaking scientific exploration, particularly due to the presence of a young naturalist named Charles Darwin onboard. Here is a detailed analysis of this landmark expedition (Figure 3.3):

Figure 3.3. HMS beagle.

Source: *By Conrad Martens (1801–1878) – English Wikipedia (13:42, 15 October 2005. User: Dave Souza 1235x821 (73563 bytes) (HMS Beagle in the* *seaways of Tierra del Fuego, painting by Conrad Martens during the voyage of the Beagle (1831–1836), from The Illustrated Origin of Species by Charles Darwin, abridged and illustrated by Richard Leakey), Public Domain, https://commons.wikimedia.org/w/index.php?curid=1082238; Creative Commons License.*

1. **Purpose and Objectives:** The primary objective of the Beagle's voyage was to map the coasts of South America, essential for navigation and trade. Additionally, the expedition aimed to collect geological, biological, and ethnographic specimens, contributing to the understanding of the natural world. Darwin's role as a naturalist was pivotal, although the voyage's scientific significance far surpassed the initial objectives.
2. **Key Destinations and Discoveries:**

i. **South America:** The Beagle extensively explored the coasts of South America, including Brazil, Argentina, Chile, and the Galapagos Islands. Darwin's observations of diverse landscapes, geological formations, and flora and fauna were foundational to his later work.

ii. **Galapagos Islands:** The Galapagos archipelago held a special significance. Darwin's observations of distinct but closely related species on different islands formed the basis for his theory of evolution by natural selection. This experience profoundly influenced his understanding of adaptation and speciation.

iii. **Australia:** The Beagle also explored the eastern coast of Australia, contributing to the mapping of this continent. Darwin's observations here, especially in places like the Great Barrier Reef, added to his understanding of geological formations and marine life.

iv. **Cape Verde Islands and Ascension Island:** The Beagle visited these islands, providing Darwin with additional opportunities to study unique ecosystems and geological features.

3. **Darwin's Observations and Collections:** Darwin made meticulous observations during the voyage, documenting various species, geological formations, and fossils. He collected specimens, including plants, animals, and geological samples, which later contributed significantly to his research. His observations of adaptations and variations among species laid the groundwork for his theory of evolution.

4. **Influence on Darwin's Theory of Evolution:** The wealth of observations and specimens Darwin gathered during the Beagle's voyage profoundly influenced his thinking. His encounters with diverse environments and species, especially in the Galapagos Islands, challenged prevailing notions of the fixity of species. Darwin's exposure to different ecosystems and his comparative studies were crucial in the development of his groundbreaking theory of evolution by natural selection.
5. **Legacy and Scientific Impact:** The scientific legacy of the Beagle's voyage is immeasurable. Darwin's experiences and collections provided the foundation for his masterpiece, "On the Origin of Species," published in 1859. The voyage's data and specimens continue to be studied, providing valuable insights into evolution, geology, and ecology. The HMS Beagle expedition remains a beacon in the history of science, illustrating the power of exploration and observation in shaping our understanding of the natural world.

3.3. ON THE ORIGIN OF SPECIES

Published in 1859, Charles Darwin's "On the Origin of Species" is one of the most influential scientific works ever written. This groundbreaking book revolutionized the understanding of life on Earth and laid the foundation for the theory of evolution by natural selection. Here's a detailed analysis of its key components and lasting impact (Figure 3.4):

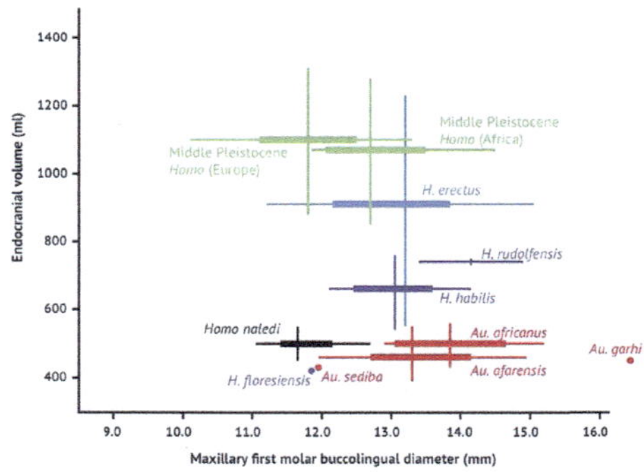

Figure 3.4. Origin of species.

Source: *By Lee Roger Berger research team – http://elifesciences.org/content/4/e09560, CC BY-SA 4.0, https://commons.wikimedia.org/w/index.php?curid=43346102; Creative Commons License.*

1. **Central Thesis: Evolution by Natural Selection:** The core concept of Darwin's book is the theory of evolution by natural selection. Darwin argued that species evolve over time through a process of gradual modification and adaptation to their environment. Those individuals with advantageous traits are more likely to survive and reproduce, passing on these traits to their offspring. This mechanism, natural selection, leads to the gradual divergence and speciation of organisms.
2. **Variation and Adaptation:** Darwin extensively discussed the variation present within species. He observed that individuals within a population exhibit variation in traits, a phenomenon crucial for the process of natural selection. Certain variations provide advantages in specific environments, leading to adaptation. The book highlights various examples of adaptation, emphasizing the role of the environment in shaping traits.
3. **Fossil Record and Transitional Forms:** Darwin addressed the fossil record, acknowledging the imperfections and gaps. He explained these gaps as due to the incomplete nature of the geological record. Darwin also discussed the concept of transitional forms, organisms displaying characteristics of both ancestral and descendant groups. He argued that these transitional forms would have existed but might not yet have been discovered.
4. **Biogeography and Geographic Distribution:** Darwin explored the geographical distribution of species, noting similarities and differences between organisms from different regions. He proposed that these patterns could be explained by the migration and adaptation of species to their local environments. This observation supported his theory of descent with modification.
5. **Artificial Selection and Natural Selection:** Darwin drew parallels between artificial selection, where humans selectively breed plants and animals for specific traits, and natural selection. He argued that if humans could produce significant changes in domestic species over a short time, nature could achieve similar results over vast geological periods through the process of natural selection.
6. **Reproduction and Heredity:** Darwin discussed the principles of reproduction and heredity, emphasizing the role of inheritance in the transmission of traits. He proposed that traits advantageous for

survival would be passed down to future generations, gradually leading to the evolution of new species.

7. **Challenges and Responses:** Darwin anticipated and addressed potential challenges to his theory. He discussed counterarguments, including the complexity of the eye and the absence of transitional forms, providing plausible explanations. His responses showcased the depth of his understanding and the robustness of his theory.
8. **Impact and Legacy:** "On the Origin of Species" had a profound impact on biology, challenging traditional views of creation and sparking intense debates. Its influence extended beyond science, shaping philosophical, religious, and ethical discussions. The book laid the foundation for modern evolutionary biology, inspiring generations of scientists to explore life's diversity and origins. Darwin's ideas continue to be central to our understanding of the natural world, underscoring the power of observation, evidence, and reasoning in scientific inquiry.

3.4. NATURAL SELECTION AS THE MECHANISM OF EVOLUTION

Natural selection, proposed by Charles Darwin as the driving force behind the process of evolution, is a fundamental concept in biology. Here's a comprehensive analysis of natural selection as the mechanism of evolution (Figure 3.5):

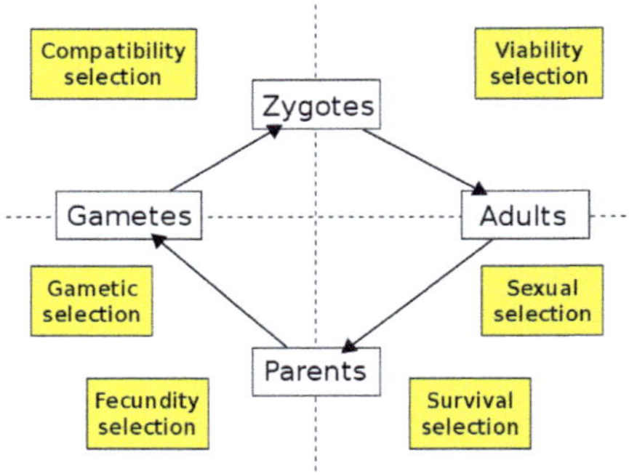

Figure 3.5. Natural selection as the mechanism of evolution.

Source: *By Wykis/Wykis on Wikipedia – Own work, based on content of w:en:File:Selection forces.gif, Public Domain, https://commons.wikimedia.org/w/index.php?curid=2064641 om; Creative Commons License.*

1. **Variation Within Populations:** Natural selection starts with the inherent variation within populations. Individuals within a species exhibit differences in traits, such as size, color, or behavior. These variations can be heritable, passed from one generation to the next through genetic inheritance.
2. **Struggle for Existence:** In any environment, resources are limited, leading to competition among individuals for survival. This competition can take various forms, including competition for food, mates, or shelter. Not all individuals within a population can survive and reproduce at their full potential due to these limitations, leading to what Darwin termed the "struggle for existence."
3. **Differential Reproduction:** Because of the variations among individuals, certain traits provide advantages in the struggle for existence. Individuals with advantageous traits are more likely to survive and reproduce, while those with less favorable traits might not survive to reproductive age or might have fewer offspring. This discrepancy in reproductive success, based on traits that enhance survival and reproduction, is at the core of natural selection.
4. **Heritability of Traits:** Crucially, the advantageous traits that lead to higher reproductive success must be heritable, meaning they can be passed on to offspring. If a particular trait that provides an advantage is heritable, it will become more prevalent in the population over successive generations. This is the essence of natural selection – the differential survival and reproduction of individuals based on heritable variations.
5. **Accumulation of Adaptations:** Over long periods of time, natural selection leads to the accumulation of adaptations in populations. Adaptations are traits or characteristics that increase an organism's fitness in its environment. As advantageous traits become more common in a population, the population becomes better suited to its environment, reflecting the gradual process of adaptation through natural selection.
6. **Role of Environmental Change:** Natural selection is dynamic and responsive to environmental changes. When the environment changes, certain traits that were advantageous might become less so, leading to the selection of new traits that are better suited to

the altered environment. This process allows populations to adapt to changing conditions over time.

7. **Examples and Observations:** Natural selection is supported by numerous examples and observations in the natural world, including the famous case of the peppered moths during the Industrial Revolution and the evolution of antibiotic resistance in bacteria. These real-world examples demonstrate the impact of natural selection on populations and how it leads to observable changes in traits over relatively short periods.

Thus, natural selection is a fundamental mechanism of evolution, explaining how populations change and adapt to their environments over generations. This elegant concept, rooted in observation and evidence, forms the cornerstone of modern evolutionary biology, shaping our understanding of the diversity of life on Earth.

3.5. HISTORY OF PALEONTOLOGY

Paleontology, the scientific study of life that existed prior to the present, has a rich and fascinating history spanning centuries. Here is a detailed analysis of the evolution of paleontology through time (Figure 3.6):

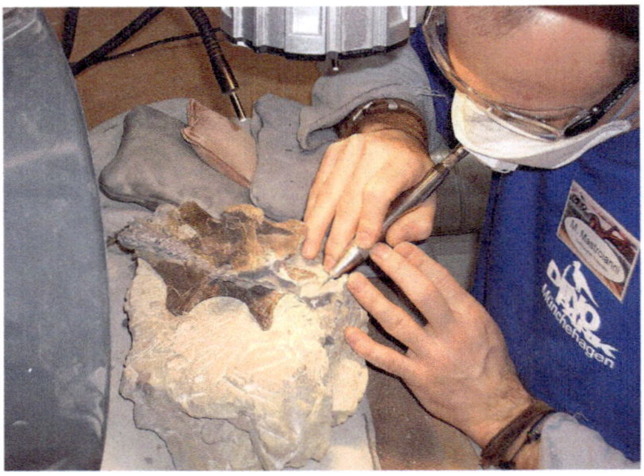

Figure 3.6. Paleontology.

Source: *By Nils Knötschke – Nils Knötschke, CC BY-SA 2.5, https://commons.wikimedia.org/w/index.php?curid=1228456; Creative Commons License.*

1. **Antiquity and Early Curiosities:**

I. **Ancient Cultures:** Fossils and ancient remains have fascinated humans for millennia. Ancient civilizations, such as the Greeks and Romans, occasionally encountered fossils and crafted mythological explanations for their existence.

II. **Chinese Fossils:** In ancient China, fossils were sometimes mistaken for dragon bones and were used in traditional medicine. These fossils often came from extinct animals, including dinosaurs.

2. **The Renaissance and the Birth of Scientific Inquiry:**

i. **Rediscovery of Fossils:** During the Renaissance in Europe, fossils were rediscovered, leading to early attempts at classification. Scholars like Leonardo da Vinci recognized the organic origin of fossils.

ii. **Emergence of Modern Ideas:** In the 17th century, scientists like Robert Hooke and Nicholas Steno made significant contributions, including the concept of stratigraphy (study of rock layers), laying the groundwork for future geological and paleontological studies.

3. 18th and 19th Centuries – Age of Discovery:

I. **Cuvier and Extinction:** Georges Cuvier, a French naturalist, proposed the concept of extinction, challenging the prevailing belief in the fixity of species. He demonstrated the existence of ancient animals, like mammoths, which no longer existed.

II. **Mary Anning:** Anning, an English fossil collector, made several significant discoveries in the early 19th century, including the first complete Ichthyosaurus skeleton. Her work contributed to the understanding of extinct marine reptiles.

III. **Evolutionary Ideas:** The 19th century saw a growing acceptance of the idea of evolution. Scientists like Charles Darwin and Alfred Russel Wallace, influenced by geological and paleontological findings, proposed the theory of evolution by natural selection.

4. 20th Century – Advancements and Specializations:

I. **Dinosaur Renaissance:** In the mid-20th century, paleontology experienced a "dinosaur renaissance" marked by extensive discoveries, leading to a deeper understanding of dinosaur biology, behavior, and evolution.

II. **Technological Advances:** Technological innovations, such as radiometric dating and electron microscopy, revolutionized paleontological research, enabling precise dating of fossils and detailed analysis of microscopic structures.

III. **Molecular Paleontology:** The study of ancient DNA became possible, allowing scientists to gain insights into the genetic relationships between extinct species and their modern relatives.

5. 21st Century – Interdisciplinary Approach:

I. **Synthesis with Other Sciences:** Paleontology increasingly collaborates with fields like genetics, geology, and climatology. These interdisciplinary approaches provide a more holistic understanding of ancient life and its interactions with the environment.

II. **Digital Paleontology:** Advanced imaging techniques and computer modeling have facilitated the study of fossils without damaging them, allowing for detailed analysis and preservation of delicate specimens.

III. **Climate Change and Extinction:** Paleontology plays a crucial role in understanding past mass extinctions and their relation to climate change. Studying ancient extinctions informs current discussions about biodiversity loss and climate-driven threats to species.

Hence, the history of paleontology is a testament to human curiosity and the persistent quest to understand the history of life on Earth. From early myths to the cutting-edge scientific methods of the 21st century, paleontology continues to uncover the mysteries of ancient life, contributing valuable insights to our knowledge of evolution, biodiversity, and the planet's geological history.

3.6. TRANSITIONAL FOSSIL

Figure 3.7. Transitional fossil.

Source: *By H. Raab (User: Vesta) – Own work, CC BY-SA 3.0, https://commons.wikimedia.org/w/index.php?curid=8066320; Creative Commons License.*

Transitional fossils are a crucial component of the fossil record, serving as tangible evidence of evolutionary change over geological time. These fossils bridge the gaps between different species, providing glimpses into the evolutionary pathways and transformations that have occurred over millions of years. Here's a detailed analysis of transitional fossils and their significance in the study of evolution (Figure 3.7):

1. **Definition and Characteristics:**

I. **Transitional Fossils:** Also known as intermediary or intermediate fossils, are the remains of organisms that exhibit traits characteristic of both ancestral and descendant groups. These fossils often possess a mix of features, indicating a transitional stage in evolutionary history.

II. **Mosaic Traits:** Transitional fossils typically display mosaic traits, where certain anatomical features resemble those of ancestral species, while others resemble features found in later, more derived species.

2. **Importance of Transitional Fossils:**

I. **Filling Evolutionary Gaps:** One of the primary roles of transitional fossils is filling gaps in the evolutionary lineage. They provide direct evidence of the gradual changes that have occurred within lineages, offering a detailed picture of the evolutionary process.

II. **Confirmation of Evolutionary Relationships:** Transitional fossils help confirm the evolutionary relationships proposed by comparative anatomy and molecular biology. They provide tangible evidence supporting the relatedness of different groups of organisms.

III. **Understanding Major Evolutionary Events:** Transitional fossils shed light on significant evolutionary events, such as the transition from fish to tetrapods, reptiles to birds, and land mammals to marine mammals. By studying these fossils, scientists can reconstruct the steps involved in these key evolutionary transitions.

3. **Notable Examples of Transitional Fossils:**

I. **Archaeopteryx:** Often considered a transitional form between reptiles (dinosaurs) and birds, Archaeopteryx had feathered wings like birds but also retained dinosaur-like traits, including teeth and a long bony tail (Figure 3.7 is of this fossil).

II. **Tiktaalik:** This fossil represents a transitional form between fish and tetrapods. It had fish-like features such as scales and gills but also tetrapod-like traits like limb bones with wrists, indicating its adaptation to both aquatic and terrestrial environments.

III. **Ambulocetus:** It is an intermediate fossil showing features of both land-dwelling mammals and modern whales. It had limbs capable of supporting its weight on land and specialized ear bones found in whales.

4. **Challenges and Controversies:**

I. **Incomplete Fossil Record:** The fossil record is inherently incomplete, making it challenging to find transitional fossils representing every evolutionary step. Some transitions might be poorly represented due to the rarity of certain organisms or conditions conducive to fossilization.

II. **Definition Debate:** There can be debates about what constitutes a true transitional fossil, with some arguing that nearly all fossils are transitional to some extent, given the continuous nature of evolution.

5. **Implications for Evolutionary Theory:**

I. **Supporting Evolution:** Transitional fossils provide empirical evidence supporting the theory of evolution. Their discovery and analysis validate the predictions made by evolutionary biology about the existence of intermediate forms connecting different groups of organisms.

II. **Continuous Change:** The presence of transitional fossils underscores the continuous nature of evolutionary change. It highlights that species are not fixed entities but are constantly evolving and adapting to changing environments over geological time.

In fact, transitional fossils play a pivotal role in our understanding of evolutionary processes. By offering glimpses into the past and illuminating the gradual transformations that have shaped life on Earth, these fossils provide tangible evidence of the interconnectedness of all living beings and the dynamic nature of evolution.

3.7. BLENDING INHERITANCE

Blending inheritance was a prominent theory in the history of genetics that suggested that hereditary traits from parents blend together in offspring, leading to a uniform intermediate trait. This concept was widely accepted before the advent of Mendelian genetics. Here's a detailed analysis of blending inheritance, its historical context, and its place in the modern understanding of genetics (Figure 3.8):

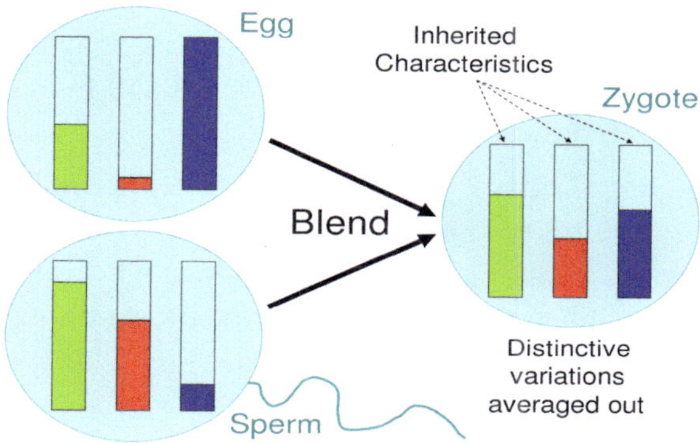

Figure 3.8. Blending inheritance.

Source: *By Ian Alexander – Own work, CC BY-SA 4.0, https://commons.wikimedia.org/w/index.php?curid=62643135; Creative Commons License.*

1. **Historical Context:** Pre-Mendelian Beliefs: Before Gregor Mendel's experiments with pea plants in the mid-19th century, the prevailing notion was that hereditary traits from parents mixed in a manner similar to how colors blend in paint. According to this theory, the traits of parents would blend in their offspring, resulting in a uniform intermediate phenotype.
2. **Challenges to Blending Inheritance:** Recessive and Dominant Traits: Mendel's experiments, which focused on traits with distinct, observable differences (such as yellow vs. green peas), revealed patterns of inheritance that contradicted blending inheritance. Mendel discovered that traits were inherited as discrete units, which he termed "factors" and are now known as genes. This explained the reappearance of traits in later generations, even if they were not visible in the immediate offspring.
3. **Modern Understanding and Genetic Principles:**

I. **Mendelian Genetics:** Mendel's laws of segregation and independent assortment laid the foundation for modern genetics. The principle of segregation states that individuals have two alleles for each gene, one inherited from each parent, and these alleles segregate during the formation of gametes. Independent assortment explains how genes located on different chromosomes segregate independently during gamete formation.

II. **Genetic Variation:** Mendelian genetics allows for the preservation of genetic variation in populations. Different combinations of alleles result in diverse phenotypes, and this variation is essential for the adaptation and evolution of species in changing environments.

III. **Polygenic Traits:** While Mendel's experiments focused on single gene traits, many traits, such as height or intelligence, are polygenic, meaning they are influenced by multiple genes. In polygenic inheritance, the interaction of multiple genes leads to a wide range of phenotypic variation.

4. **Legacy of Blending Inheritance:**

I. **Misconceptions and Misinterpretations:** Despite the acceptance of Mendelian genetics, some misconceptions rooted in blending inheritance persist. People may still occasionally think of traits as blending together, especially in cases where the contributions of multiple genes are not fully understood.

II. **Educational Significance:** Understanding the historical context of blending inheritance is crucial for appreciating the revolutionary impact of Mendel's work. It highlights the shift from vague, non-specific ideas about inheritance to the precise and testable principles of Mendelian genetics.

Blending Inheritance as a Milestone: While blending inheritance was an important step in the history of genetics, the shift to Mendelian principles marked a profound advancement. The understanding of heredity transformed from a vague concept of blending traits to a precise science, providing the basis for modern genetics and our understanding of the complexities of inheritance and variation.

REFERENCES

1. Charles, D. *On the Origin of Species.*
2. Charles, D. *The Voyage of the Beagle.*
3. David, Q. *The Reluctant Mr. Darwin: An Intimate Portrait of Charles Darwin and the Making of His Theory of Evolution.*
4. Iain, M. *Darwin's Armada: Four Voyages and the Battle for the Theory of Evolution.*
5. Janet, B. *Charles Darwin: The Power of Place.*
6. Janet, B. *Charles Darwin: Voyaging.*
7. Janet, B. *Darwin's Origin of Species: A Biography.*
8. Michael, W. *Darwin: A Life in Science.*
9. Rebecca, S. *Darwin's Ghosts: The Secret History of Evolution.*
10. Richard, D. K. *Charles Darwin's Beagle Diary.*

CHAPTER 4

Darwin's Contemporaries and Alternative Theories

CONTENTS

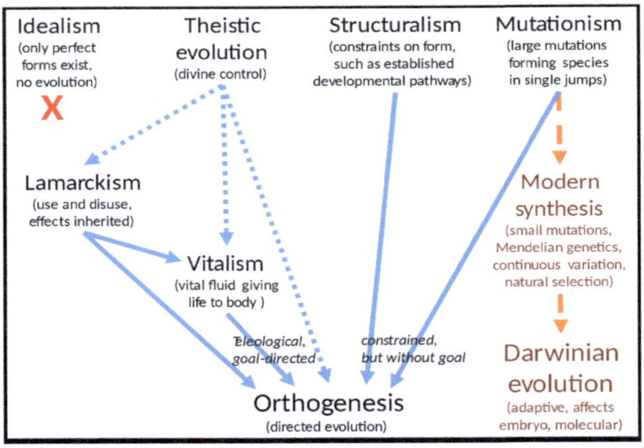

Figure 4.1. Darwin's contemporaries and alternative theories.

Source: *By Ian Alexander – Own work, CC BY-SA 4.0, https://commons.wikimedia.org/w/index.php?curid=62298929; Creative Commons License.*

4.1. ALFRED RUSSEL WALLACE AND INDEPENDENT DISCOVERY

Alfred Russel Wallace's independent discovery of the theory of evolution by natural selection stands as one of the most remarkable events in the history of science. Here's a detailed analysis of Alfred Russel Wallace's independent discovery and its profound implications (Figure 4.2):

Figure 4.2. Alfred Russel Wallace.

Source: *oumnh.ox.ac.uk; Creative Commons License.*

1. **Early Life and Passion for Natural History:**

I. **Self-Taught Naturalist:** Wallace, born in 1823 in Wales, was largely self-taught in the field of natural history. His curiosity about the natural world led him to extensive travels and observations, particularly in South America and Southeast Asia.

2. **Explorations and Scientific Contributions:**

I. **Amazon Expeditions:** During his expedition to the Amazon rainforest, Wallace collected an enormous number of specimens and noted significant biogeographical patterns. He observed the diversity of species and became intrigued by the variations and adaptations in different environments.

II. **Malay Archipelago:** Wallace's explorations in the Malay Archipelago, where he spent eight years, were particularly influential. His observations of geographical distribution and species variation played a pivotal role in shaping his evolutionary ideas.

3. **Independent Formulation of Evolutionary Theory:**

I. **Natural Selection Concept:** Wallace, independently of Charles Darwin, conceived the idea of natural selection as the mechanism driving evolution. He realized that species adapted to their environments, leading to the survival of the fittest.

II. **Letter to Darwin:** In 1858, while suffering from malaria in the Malay Archipelago, Wallace wrote a letter to Charles Darwin outlining his theory of natural selection. This letter, sent to Darwin, profoundly shocked him as it mirrored his own unpublished thoughts on evolution.

4. Joint Presentation with Darwin:

I. **Linnean Society Meeting:** Darwin, recognizing the significance of Wallace's work, arranged for a joint presentation of his own ideas and Wallace's paper at the Linnean Society of London in 1858. This event marked the public introduction of the theory of evolution by natural selection.

II. **Publication of "On the Origin of Species":** Darwin, spurred by Wallace's letter, hastened the publication of his monumental work, "On the Origin of Species," in 1859. The book presented a comprehensive exposition of the theory, incorporating years of Darwin's research and observations.

5. **Legacy and Recognition:**

I. **Scientific Recognition:** Wallace's contributions to evolutionary biology were widely recognized during his lifetime. He received numerous awards and honors, including the Copley Medal from the Royal Society.

II. **Legacy in Modern Biology:** Wallace's work laid the foundation for modern evolutionary biology. His emphasis on natural selection as the driving force of evolution remains a cornerstone of the field, and he is remembered as one of the key figures in the theory of evolution.

In summary, Alfred Russel Wallace's independent discovery of the theory of evolution by natural selection, alongside Darwin, underscores the simultaneous intellectual exploration happening in the 19th century. It exemplifies the power of keen observation, critical thinking, and scientific intuition that led to one of the most transformative ideas in the history of biology. Wallace's contributions continue to be celebrated, reminding us of the collaborative and dynamic nature of scientific progress.

4.2. CHALLENGES AND CRITICS OF DARWIN'S IDEAS

Charles Darwin's theory of evolution by natural selection revolutionized the understanding of life on Earth. However, it was met with significant challenges and criticisms during his time and continues to face scrutiny even today. Here's a detailed analysis of the challenges and critics faced by Darwin's ideas:

1. **Religious Opposition:**

I. **Conflict with Creationism:** Darwin's theory contradicted the biblical account of creation, leading to strong opposition from religious communities. Many religious leaders saw evolution as a direct challenge to the notion of divine creation.

II. **Theological Implications:** The idea that species evolved through natural processes challenged the traditional view of a purposeful, unchanging creation, raising theological questions about the existence of God's design in the natural world.

2. **Scientific Criticisms:**

I. **Lack of Mechanistic Understanding:** Critics argued that Darwin's theory lacked a detailed explanation of the mechanisms driving evolution. The understanding of genetics and heredity, which provides the foundation for modern evolutionary biology, was not available in Darwin's time.

II. **Origin of Complex Structures:** Skeptics questioned how complex organs and structures, such as the eye, could have evolved gradually through natural selection, considering the intricate nature of these adaptations.

3. **Paleontological Challenges:** Gaps in the Fossil Record: Critics pointed out gaps in the fossil record, emphasizing the absence of transitional forms (intermediate species) between major groups of organisms. These gaps were seen as a challenge to the gradualism proposed by Darwin.

4. **Social and Ethical Concerns:**

I. **Social Darwinism:** Darwin's ideas were misapplied to human societies, leading to the emergence of social Darwinism. This concept, advocating the survival of the fittest in human societies, had significant ethical implications and was used to justify various forms of discrimination.

II. **Eugenics Movements:** The misuse of evolutionary concepts also contributed to the rise of eugenics movements, leading to unethical practices such as forced sterilization in the early 20th century.

5. Modern Challenges:

I. **Intelligent Design and Creationism:** In the modern era, challenges to Darwin's ideas come in the form of intelligent design and creationism. Proponents of intelligent design argue that certain biological features are too complex to have evolved and must be the result of an intelligent designer.

II. **Antievolution Legislation:** Efforts to introduce antievolution legislation in schools persist, reflecting ongoing challenges to the teaching of evolution in some educational contexts.

6. Scientific Advances and Responses:

I. **Genetics and Molecular Biology:** Advances in genetics and molecular biology have provided extensive evidence supporting evolutionary theory. The discovery of DNA and the understanding of genetic mutations have bolstered the understanding of heredity and variation.

II. **Transitional Fossils:** Ongoing paleontological discoveries continue to fill gaps in the fossil record, revealing numerous transitional fossils that illustrate the gradual changes between species over time.

III. **Experimental Evolution:** Experimental studies, including the observation of evolutionary processes in real-time in controlled environments, provide empirical evidence supporting Darwinian evolution.

So, while Darwin's ideas faced and continue to face challenges, the overwhelming body of scientific evidence from various disciplines supports the theory of evolution by natural selection. Scientific advancements, coupled with ongoing research, have addressed many of the initial criticisms and solidified the theory as the foundation of modern biology. The challenges to Darwin's ideas highlight the importance of critical thinking and ongoing scientific inquiry in the pursuit of understanding the complexities of life's evolutionary history.

4.3. OTHER EVOLUTIONARY THEORIES OF THE TIME

During the 19th century, when Charles Darwin was formulating his theory of evolution by natural selection, there were several other evolutionary theories and ideas proposed by scientists and thinkers. Here are some notable evolutionary theories of the time:

1. **Lamarckism:** Proposed by Jean-Baptiste Lamarck.
   - **Key Ideas:** Lamarckism posited that organisms could change during their lifetimes in response to their environment, and these acquired traits could be passed on to the next generation. This theory suggested a continuous progression toward greater complexity and perfection in organisms.
2. **Orthogenesis:**
   - **Key Ideas:** Orthogenesis, also known as straight-line evolution, suggested that species have an innate tendency to evolve in a predetermined direction. This theory implied that evolution had a built-in purpose or goal, leading species along a fixed path, regardless of natural selection.
3. **Saltationism:**
   - **Key Ideas:** Saltationism proposed that evolutionary changes occur in leaps or sudden jumps, rather than through the gradual

accumulation of small changes. Advocates of this theory believed that new species could arise suddenly, fully formed, rather than through a gradual process of small modifications.

4. **Catastrophism:** Proposed by Georges Cuvier.

- **Key Ideas:** Catastrophism suggested that the Earth's geological features were primarily the result of sudden, short-lived, violent events, such as floods or earthquakes, rather than gradual processes. While not a biological theory of evolution, it influenced early ideas about the history of life on Earth.

5. **Darwin's Theory of Evolution by Natural Selection:** Proposed by Charles Darwin.

- **Key Ideas:** Darwin's theory, as outlined in his book "On the Origin of Species" (1859), proposed that species evolve over time through the process of natural selection. Variation within species leads to differential survival and reproduction, with advantageous traits being passed on to the next generation. Over long periods, this process results in the adaptation of organisms to their environments and the emergence of new species.

6. **Mutation Theory:** Proposed by Hugo de Vries

- **Key Ideas:** Mutation theory suggested that new species arise from sudden, large, heritable changes (mutations) in organisms. De Vries, while studying plant breeding, proposed this theory based on his observations of unusual plant variants. Mutation theory later influenced the modern understanding of genetic variation (Figure 4.3).

Figure 4.3. Hugo de Vries's mutation theory.

Source: *By Thérèse Schwartze – Universiteitsmuseum Amsterdam, Public Domain, https://commons.wikimedia.org/w/index.php?curid=20440556; Creative Commons License.*

7. **Biometricians and Statistical Darwinism:**

- **Key Ideas:** Biometricians, led by Francis Galton, believed that Darwin's theory could be supported using statistical methods and the study of continuous traits (such as height or weight). They focused on the statistical analysis of variation within populations, emphasizing the role of small, continuous changes (Figure 4.4).

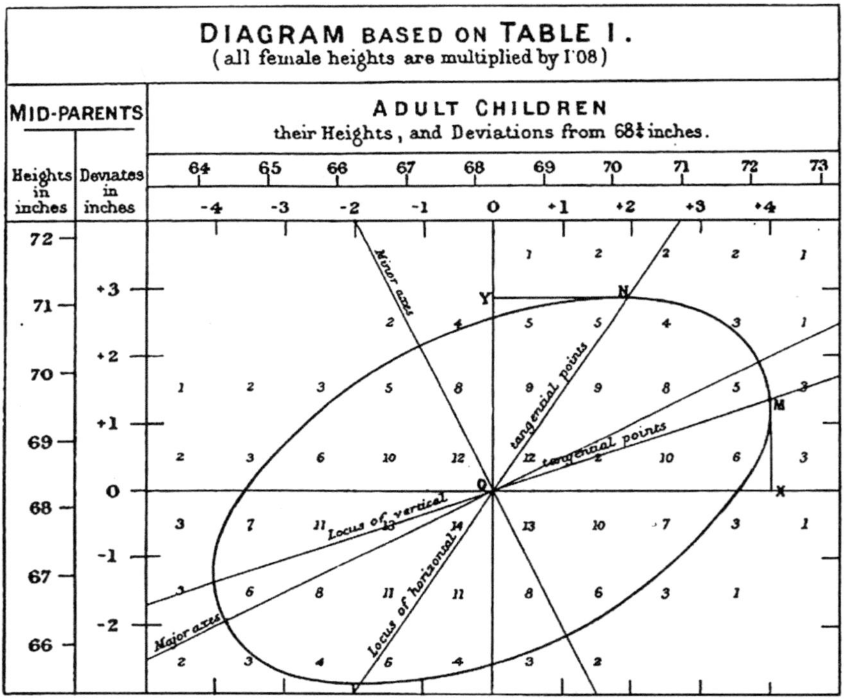

Figure 4.4. Francis Galton's biometric sketches.

Source: *By Francis Galton – http://euclid.psych.yorku.ca/SCS/Gallery/images/galton-corr.jpg, Public Domain, https://commons.wikimedia.org/w/index.php?curid=29526492; Creative Commons License.*

However, it's important to note that Darwin's theory of evolution by natural selection eventually became the dominant and widely accepted explanation for the diversity of life on Earth due to its explanatory power, backed by extensive evidence from various scientific fields. Many of the alternative theories of the time were gradually overshadowed as the understanding of genetics, heredity, and the fossil record advanced.

REFERENCES

1. Alpheus, S. P. *Lamarck, The Founder of Evolution: His Life and Work.*
2. Barbara, F., & Paul, R. G. *Creationism's Trojan Horse: The Wedge of Intelligent Design.*
3. Jean, G. *The Transformist Illusion.*
4. Keith, T. *Before Darwin: Reconciling God and Nature.*
5. Martin, J. S. R. *Georges Cuvier, Fossil Bones, and Geological Catastrophes: New Translations and Interpretations of the Primary Texts.*
6. Paul, C. *Darwinism, War, and History: The Debate Over the Biology of War from the 'Origin of Species' to the First World War.*
7. Peter, J. B. *Evolution: The History of an Idea.*
8. Peter, J. B. *The Eclipse of Darwinism: Anti-Darwinian Evolution Theories in the Decades around 1900.*
9. Robert, C. *Vestiges of the Natural History of Creation.*
10. Robert, J. R. *Darwin and the Emergence of Evolutionary Theories of Mind and Behavior.*

CHAPTER 5

The Modern Synthesis

CONTENTS

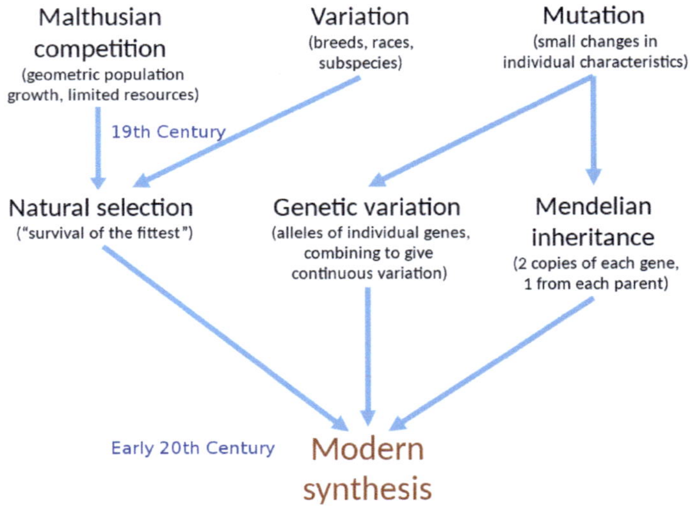

Figure 5.1. The modern synthesis in evolution.

Source: *By Ian Alexander – Own work, CC BY-SA 4.0, https://commons.wikimedia.org/w/index.php?curid=62385744; Creative Commons License.*

5.1. MENDELIAN GENETICS AND EVOLUTION

Mendelian genetics, based on the principles established by Gregor Mendel, plays a fundamental role in understanding the mechanisms of evolution. Here's an in-depth analysis of Mendelian genetics and its intricate connection to the process of evolution:

1. **Mendel's Laws:**

I. **Law of Segregation:** Mendel's first law states that each individual has two alleles for each gene, one inherited from each parent. These alleles segregate (separate) during the formation of gametes, ensuring that each gamete carries only one allele for each gene.

II. **Law of Independent Assortment:** Mendel's second law states that the alleles of different genes segregate, or assort, independently of one another during gamete formation. This means that the inheritance of an allele for one gene does not affect the inheritance of an allele for another gene, allowing for a wide array of genetic combinations.

2. **Genetic Variability and Evolution:**

I. **Mutation:** Which are changes in DNA sequences, provide the raw material for evolution. Mutations can create new alleles, introducing genetic variation within populations.

II. **Recombination:** Genetic recombination during meiosis results in new combinations of alleles in offspring. This shuffling of genetic material increases genetic diversity within populations.

III. **Sexual Reproduction:** By combining genetic material from two parents, increases genetic variability among individuals in a population. This diversity is essential for natural selection to act upon, enabling populations to adapt to changing environments.

3. **Population Genetics and Evolutionary Forces:**

I. **Natural Selection:** It acts on the variation present within populations. Certain alleles confer advantages, enhancing the organism's fitness and survival. Over generations, advantageous alleles become more prevalent in a population, leading to adaptive evolution.

II. **Genetic Drift:** The random change in allele frequencies over generations, occurs in small populations due to chance events. Genetic drift is a significant evolutionary force, particularly in small or isolated populations, leading to the fixation or loss of alleles.

III. **Gene Flow:** The movement of alleles between populations through migration and interbreeding, can homogenize populations genetically. However, it can also introduce new alleles, increasing genetic diversity within populations.

IV. **Mutation and Migration:** Mutation introduces new genetic variation, while migration (gene flow) transfers genetic diversity between populations. Together, these processes contribute to the adaptive potential of populations over time.

4. **Speciation and Mendelian Inheritance:**

I. **Allopatric Speciation:** Geographic isolation can lead to allopatric speciation, where populations are separated by a physical barrier. Over time, genetic differences accumulate due to mutation, genetic drift, and natural selection, leading to the formation of distinct species.

II. **Sympatric Speciation:** It occurs without physical separation. It can be driven by factors such as polyploidy (an increase in chromosome number), disruptive selection, or assortative mating, leading to the divergence of populations within the same geographical area.

5. **Selective Pressures and Adaptation:**

i. **Adaptation:** Mendelian genetics underpins the concept of adaptation, where advantageous traits increase an organism's fitness in a specific environment. Natural selection acts on these traits, leading to the evolution of populations that are well-suited to their habitats.

ii. **Convergent Evolution:** It occurs when different species independently evolve similar traits in response to similar selective pressures. Mendelian genetics explains how these traits are inherited and can spread within populations.

6. **Human Evolution and Mendelian Inheritance:**

i. **Human Genetic Variation:** Mendelian genetics explains the inheritance patterns of traits in humans. Understanding the genetics of diseases, behaviors, and physiological traits in humans is crucial for studying human evolution and the genetic basis of diseases.

ii. **Population Genetics in Human Evolution:** Studying the genetic variation within human populations using Mendelian principles provides insights into human evolutionary history, migration patterns, and adaptation to diverse environments (Figure 5.2).

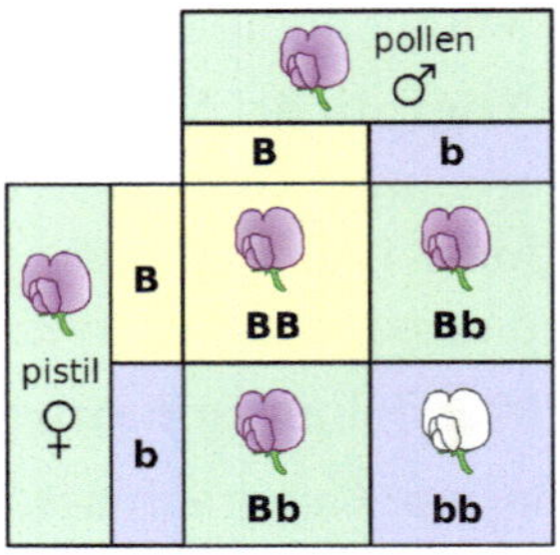

Figure 5.2. The Mendelian genetics.

Source: *By Madprime – Own work, CC BY-SA 3.0, https://commons.wikimedia.org/w/index.php?curid=2063426; Creative Commons License.*

In summary, Mendelian genetics provides the framework for understanding genetic variation within populations, the processes of evolution, and the mechanisms underlying adaptation and speciation. The integration of Mendelian principles with the study of evolutionary processes has significantly advanced our understanding of the complexities of life's evolution and the incredible diversity of living organisms on Earth.

5.2. POPULATION GENETICS

Population genetics is a branch of genetics that studies the genetic variation within and between populations, and how this variation changes over time. It provides crucial insights into evolutionary processes, adaptation, and the genetic basis of diseases. Here's an in-depth analysis of population genetics, exploring its key concepts and applications in biological research (Figure 5.3):

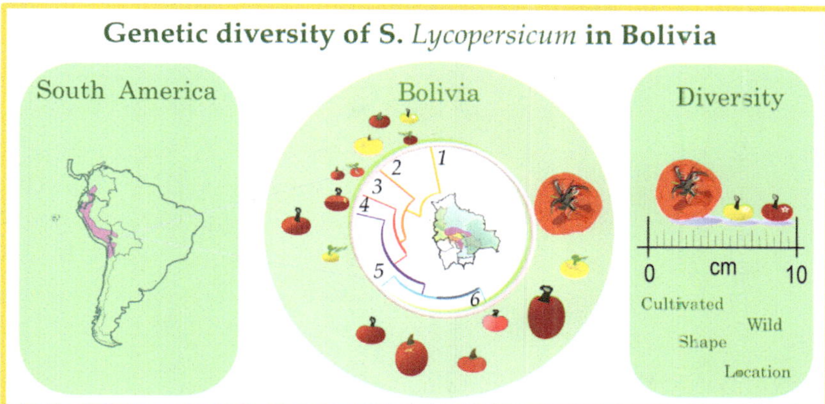

Figure 5.3. Population genetics of plant pathogens.

Source: *Villanueva-Gutierrez, E. E., Johansson, E., Prieto-Linde, M. L., Centellas, Q. A., Olsson, M. E., & Geleta, M., (2022). Simple sequence repeats markers reveal genetic diversity and population structure of Bolivian wild and cultivated tomatoes (Solanum lycopersicum L.). Genes, 13(9), 1505. https://doi.org/10.3390/genes13091505; Creative Commons License.*

1. **Genetic Variation in Populations:**

i. **Genetic Polymorphism:** Populations often exhibit genetic polymorphism, where multiple alleles exist for a particular gene. This diversity arises due to mutations, which are the ultimate source of genetic variation.

ii. **Single Nucleotide Polymorphisms (SNPs)**: These are the most common type of genetic variation, representing single base differences in DNA sequences. They are valuable markers for studying genetic traits and diseases in populations.

2. **Hardy-Weinberg Equilibrium (HWE) (Figure 5.4):**

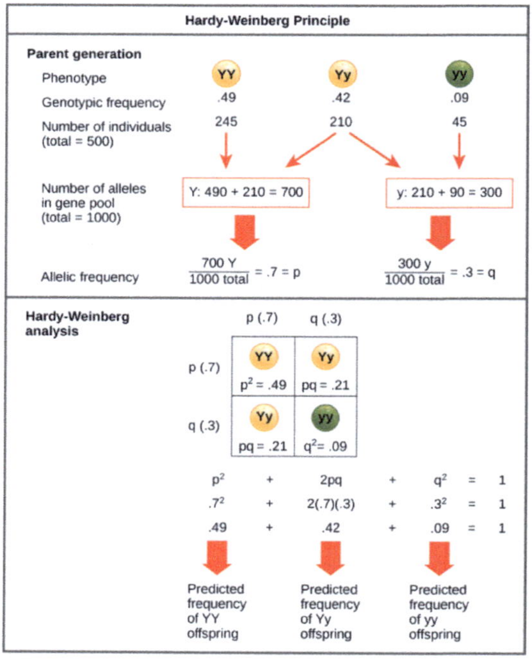

Figure 5.4. Hardy-Weinberg equilibrium (HWE).

Source: *https://bio.libretexts.org/Bookshelves/Introductory_and_General_Biology/Book%3A_General_Biology_%28Boundless%29/19%3A_The_Evolution_of_Populations/19.01%3A_Population_Evolution/19.1C%3A_Hardy-Weinberg_Principle_of_Equilibrium; Creative Commons License.*

I. **Principle:** HWE describes a theoretical population in which both the frequencies and proportions of alleles remain constant from generation to generation in the absence of other influences like mutation, selection, migration, and genetic drift.

II. **Use:** Deviations from HWE in real populations provide valuable information about the forces of evolution acting on those populations, such as detecting the presence of selection, migration, or genetic drift.

3. Genetic Drift (Figure 5.5):

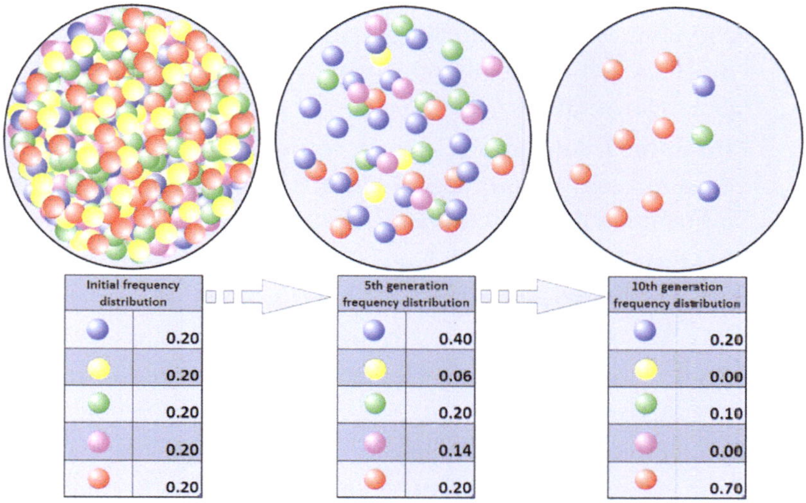

Figure 5.5. Genetic drift and gene flow.

Source: *By Professor marginalia – Own work, CC BY-SA 3.0, https://commons.wikimedia.org/w/index.php?curid=6174935; Creative Commons License.*

i. **Definition:** Genetic drift refers to random fluctuations in allele frequencies in small populations due to chance events. It is more significant in smaller populations where chance can have a pronounced effect on allele frequencies.

ii. **Bottlenecks and Founder Effects:** Genetic drift is particularly noticeable after population bottlenecks (sharp reduction in population size) and founder effects (when a small group establishes a new population), leading to reduced genetic diversity.

4. **Natural Selection:**

i. **Types:** Natural selection can be stabilizing (favoring intermediate phenotypes), directional (favoring one extreme phenotype), or disruptive (favoring both extremes). Each type shapes the genetic variation within populations based on the fitness of specific traits.

ii. **Adaptation:** Natural selection leads to adaptation, where advantageous alleles become more common, enhancing the population's fitness in its environment.

5. **Migration and Gene Flow:**

i. **Impact:** Migration introduces new alleles into populations and can counteract the effects of genetic drift, maintaining genetic diversity. It also homogenizes populations, making them genetically similar.

ii. **Role in Evolution:** Gene flow can facilitate the spread of advantageous traits or lead to the mixing of divergent populations, influencing the evolutionary trajectories of species.

6. **Mutation and Genetic Variation:**

i. **Source of Genetic Variation:** Mutations are the ultimate source of new genetic variation. While most mutations are neutral or deleterious, occasionally a beneficial mutation arises, contributing to adaptive evolution.

ii. **Molecular Clock:** Mutations in DNA sequences accumulate over time. By comparing genetic differences between species, scientists can estimate the time since their common ancestry, providing insights into evolutionary timelines.

7. **Quantitative Genetics:**

i. **Polygenic Traits:** Many traits, such as height or intelligence, are influenced by multiple genes. Quantitative genetics studies the inheritance of these complex, polygenic traits and their variation within populations.

ii. **Heritability:** It estimates the proportion of phenotypic variation in a population that is due to genetic differences. It helps understand the genetic basis of traits and their response to selection.

8. **Applications in Human Genetics:**

i. **Disease Genetics:** Population genetics plays a pivotal role in understanding the genetic basis of diseases. Genome-wide association studies (GWAS) identify genetic variants associated with diseases, enabling personalized medicine and targeted therapies.

ii. **Population History:** Studying genetic diversity in human populations provides insights into human migration patterns, historical demography, and evolutionary history, aiding anthropological and archaeological research.

9. Conservation Genetics:

i. **Genetic Diversity in Endangered Species:** Population genetics guides conservation efforts by assessing the genetic diversity within endangered species. Maintaining genetic diversity is crucial for the long-term survival and adaptability of species in changing environments.

ii. **Population Viability Analysis (PVA):** Informed by population genetics, assesses the likelihood of a population's survival over a specific period. It helps conservationists make informed decisions about managing small or declining populations.

Hence, population genetics illuminates the mechanisms driving genetic variation within populations and its impact on evolution, adaptation, and disease susceptibility. Its applications in diverse fields, from evolutionary biology to human genetics and conservation, highlight its significance in understanding the complexities of life's genetic diversity and the processes shaping the living world.

5.3. SYNTHESIZING GENETICS AND NATURAL SELECTION

The synthesis of genetics and natural selection represents a cornerstone in modern biology, providing a unified framework for understanding how genetic variation within populations is shaped by the forces of natural selection. This synthesis, often referred to as the modern synthesis or the neo-Darwinian synthesis, has profoundly shaped our understanding of evolutionary processes. Here's an in-depth exploration of the integration of genetics and natural selection:

1. **Mendelian Genetics and Heredity:**

i. **Mendel's Laws:** Gregor Mendel's laws of segregation and independent assortment laid the foundation for understanding how genetic traits are inherited from one generation to the next.

ii. **Allele Frequencies:** Mendelian genetics describes how different alleles (gene variants) are inherited and how their frequencies change within populations over time.

2. **Genetic Variation:**

I. **Sources of Variation:** Genetic variation within populations arises from mutations, recombination during meiosis, and gene flow (migration). Mutations provide the raw material for evolutionary change.

II. **Genetic Polymorphism:** Populations often exhibit genetic polymorphism, where multiple alleles exist for a given gene, leading to diversity in traits.

3. **Natural Selection:**

I. **Types of Natural Selection:** Natural selection acts on heritable traits, leading to directional selection (favoring one extreme phenotype), stabilizing selection (favoring intermediate phenotypes), or disruptive selection (favoring both extremes). These processes shape the genetic makeup of populations.

II. **Fitness:** It measures an organism's reproductive success in a specific environment. Traits that enhance an organism's fitness become more prevalent in a population over generations.

4. **Population Genetics and Evolutionary Forces:**

I. **Genetic Drift:** The random change in allele frequencies in small populations, can lead to the fixation or loss of alleles. In small populations, chance events have a significant impact on genetic diversity.

II. **Gene Flow:** This between populations introduces new alleles and can counteract the effects of genetic drift, maintaining genetic diversity within and between populations.

5. **Adaptation and Evolution:**

I. **Adaptive Evolution:** Natural selection favors traits that enhance an organism's adaptation to its environment. Over time, populations adapt to their specific ecological niches, leading to the evolution of specialized traits.

II. **Speciation:** The accumulation of genetic differences between populations, often due to adaptation to different environments, can lead to the formation of new species through a process known as speciation.

6. **Molecular Genetics and Evolution:**

I. **Genetic Basis of Traits:** Advances in molecular genetics have allowed scientists to identify specific genes and mutations underlying adaptive traits. Studying these genes provides insights into the mechanisms of evolution.

II. **Comparative Genomics:** Comparing the genomes of different species, helps identify conserved genes and regulatory elements, revealing evolutionary relationships and adaptations.

7. **Population Genomics:**

I. **Whole Genome Sequencing:** Population genomics involves sequencing the entire genomes of individuals within populations. It provides detailed insights into genetic variation, population structure, and evolutionary history.

II. **Selective Sweeps:** Population genomics identifies selective sweeps, where advantageous alleles rapidly increase in frequency due to strong positive selection, providing evidence of recent adaptation.

8. **Evolvability and Genetic Constraints:**

I. **Evolvability:** It refers to the ability of populations to generate heritable genetic variation, facilitating adaptation to new environments. Genetic mechanisms that enhance evolvability are subject to natural selection.

II. **Genetic Constraints:** Some traits are constrained by genetic factors, limiting the potential variation. These constraints influence the direction and rate of evolutionary change.

9. **Genetic Drift, Selection, and Conservation:**

I. **Conservation Genetics:** Understanding the interplay between genetic drift and natural selection is crucial in conservation genetics. Small populations are vulnerable to genetic drift, reducing genetic diversity and adaptive potential.

II. **Managing Genetic Diversity:** Conservation efforts aim to preserve genetic diversity within populations, ensuring their ability to adapt to changing environments and reducing the risk of extinction.

10. **Human Evolution and Genetic Adaptations:**

I. **Human Evolutionary Genetics:** Human populations have undergone adaptive evolution in response to diverse environments. Genetic adaptations, such as lactose tolerance in some populations, provide insights into human evolutionary history.

II. **Disease Resistance:** The study of genetic variations associated with disease resistance in human populations sheds light on the ongoing coevolutionary interactions between pathogens and human immune systems.

11. **Future Directions:**

I. **Genome Editing and Evolution:** Technologies like CRISPR-Cas9 offer the potential to directly edit genes, raising ethical questions about the deliberate manipulation of evolutionary processes.

II. **Understanding Complex Traits:** Advances in understanding polygenic traits, influenced by multiple genes and environmental factors, provide new challenges and opportunities for unraveling the genetic basis of complex diseases and behaviors.

In fact, the synthesis of genetics and natural selection represents a dynamic and ongoing exploration of the mechanisms driving evolutionary change. This integration has not only deepened our understanding of the past but also offers invaluable insights into the future of evolution, genetics, and the complex interplay between organisms and their environments. The interdisciplinary nature of this synthesis continues to drive discoveries, shaping the landscape of modern biology and our comprehension of life's remarkable diversity and adaptation.

5.4. THE NEO-DARWINIAN REVOLUTION

The Neo-Darwinian revolution, often referred to as the modern synthesis, marked a pivotal moment in the history of biology.

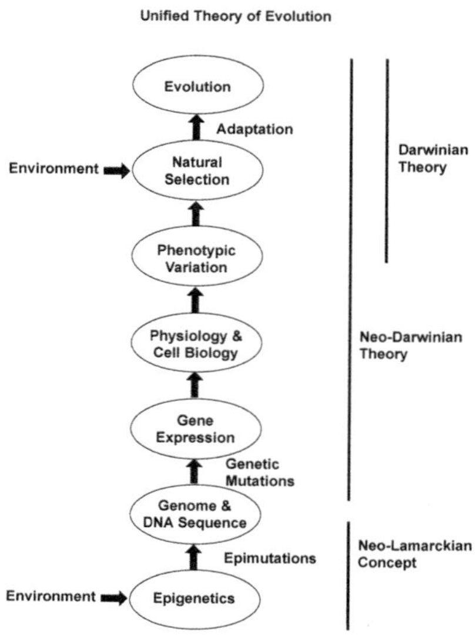

Figure 5.6. Neo-Darwinian revolution.

Source: *Nilsson, E. E., Maamar, M. B., & Skinner, M. K., (2020). Environmentally induced epigenetic transgenerational inheritance and the Weismann bar-*

rier: The Dawn of Neo-Lamarckian theory. Journal of Developmental Biology, 8(4), 28. https://doi.org/10.3390/jdb8040028; Creative Commons License.

It represents the integration of Darwinian natural selection with Mendelian genetics, leading to a unified theory of evolution. Here's an in-depth analysis of the Neo-Darwinian revolution and its profound implications for our understanding of evolution (Figure 5.6):

1. **Background and Historical Context:**

I. **Darwin's Legacy:** Charles Darwin proposed the theory of evolution by natural selection in the mid-19th century, explaining how species change over time. However, his theory lacked a clear understanding of the underlying mechanisms, especially concerning heredity.

II. **Mendelian Genetics:** Gregor Mendel's work on inheritance, published in the late 19th century, provided the framework for understanding how traits are passed from one generation to the next. Mendelian genetics introduced the concept of discrete, heritable units (genes).

2. **The Modern Synthesis:**

I. **Key Contributors:** The modern synthesis, occurring from the 1930s to the 1950s, involved prominent scientists such as Theodosius Dobzhansky, Julian Huxley, Ernst Mayr, and others. These scientists synthesized Darwinian natural selection with Mendelian genetics, bridging the gap between macroevolutionary patterns and microevolutionary processes.

II. **Integration of Genetics:** The modern synthesis unified Darwin's ideas about natural selection acting on variation with Mendel's principles of inheritance. It explained how genetic variation within populations, influenced by mutations, recombination, and gene flow, could lead to adaptive evolution through natural selection.

3. **Key Components of the Modern Synthesis:**

I. **Population Genetics:** The field of population genetics, developed during the modern synthesis, provided mathematical models to describe how genes and traits spread in populations over generations. It quantified the changes in allele frequencies due to selection, mutation, drift, and migration.

II. **Adaptation and Natural Selection:** The modern synthesis emphasized the central role of natural selection in driving adaptive evolution. It explained how advantageous traits, conferred by specific alleles, increase an organism's fitness, and become more prevalent in populations over time.

III. **Gradualism and Speciation:** The modern synthesis supported the concept of gradualism, where small, incremental changes over long periods lead to the evolution of new species. It also explained the mechanisms of speciation, including geographic isolation and reproductive barriers.

4. **Genetic Variation and Evolution:**

I. **Mutation and Genetic Diversity:** Mutations provide the raw material for evolution by creating new genetic variants. The modern synthesis incorporated the role of mutations in generating genetic diversity, which natural selection acts upon.

II. **Gene Flow and Divergence:** Gene flow between populations can homogenize them genetically, while reproductive isolation mechanisms (like geographical barriers or mating preferences) lead to divergence and speciation. The modern synthesis explained how these processes contribute to biodiversity.

5. **Implications and Applications:**

I. **Modern Evolutionary Biology:** The modern synthesis laid the foundation for modern evolutionary biology, guiding research in fields such as ecology, behavior, and molecular biology. It provided a unifying framework for understanding the diversity of life.

II. **Medicine and Agriculture:** Understanding evolutionary principles has practical applications in medicine (e.g., studying the evolution of drug resistance) and agriculture (e.g., breeding crops and livestock for desired traits).

III. **Conservation Biology:** Evolutionary concepts inform conservation strategies, such as managing genetic diversity within populations to enhance their adaptive potential in changing environments.

6. **Critiques and Expansions:**

I. **Challenges to the Synthesis:** Some critics argued that the modern synthesis was too focused on natural selection, neglecting other mechanisms like genetic drift and non-adaptive evolution. Additionally, the synthesis was criticized for not integrating developments in molecular biology.

II. **Extended Evolutionary Synthesis:** Recent advancements, including discoveries in molecular genetics and epigenetics, have led to discussions about an extended evolutionary synthesis. This new perspective considers additional factors beyond natural selection, such as developmental processes and ecological interactions.

7. **Legacy and Future Directions:**

I. **Scientific Legacy:** The modern synthesis transformed evolutionary biology into a rigorous scientific discipline, providing a framework that continues to be refined and expanded upon.

II. **Future Directions:** Current research explores the complexities of evolution, integrating genetics, genomics, and ecological interactions. Ongoing studies aim to understand the mechanisms underlying evolutionary innovations, complex traits, and the evolution of diverse biological forms.

In summary, the Neo-Darwinian revolution reshaped our understanding of evolution, unifying Darwin's natural selection with Mendelian genetics. Its legacy endures in contemporary evolutionary biology, while ongoing research pushes the boundaries of our knowledge, revealing the intricate processes that drive the incredible diversity and complexity of life on Earth.

REFERENCES

1. Douglas, J. F. *Evolutionary Biology.*
2. Ernst, M., & William, B. P. *The Evolutionary Synthesis: Perspectives on the Unification of Biology.*
3. George, C. W. *Adaptation and Natural Selection.*
4. Julian, H. *Evolution: The Modern Synthesis.*
5. Richard, D. *The Extended Phenotype.*
6. Richard, D. *The Selfish Gene.*
7. Robert, H. M., & Edward, O. W. *The Theory of Island Biogeography.*
8. Ronald, A. F. *The Genetical Theory of Natural Selection.*
9. Stephen, J. G. *Evolution as Fact and Theory.*
10. Stephen, J. G. *The Structure of Evolutionary Theory.*
11. Theodosius, D. *Genetics and the Origin of Species.*
12. William, B. P. *Theodosius Dobzhansky: A Life in Science.*

CHAPTER 6

Evidence for Evolution

CONTENTS

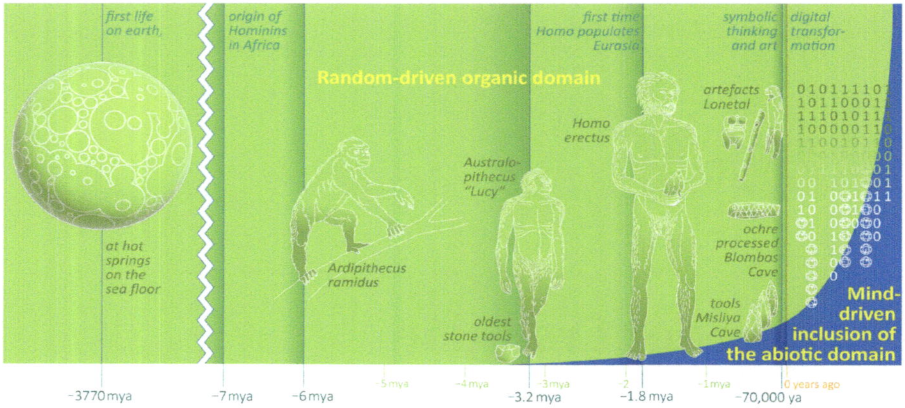

Figure 6.1. Evidence for evolution.

Source: *Weber, G. W., (2023). Quantum leaps in human biocultural evolution and the relationship to cranial capacity. Life, 13(4), 1030. https://doi.org/10.3390/life13041030; Creative Commons License.*

6.1. FOSSIL RECORD AND TRANSITIONAL FORMS

The fossil record stands as one of the most compelling pieces of evidence supporting the theory of evolution. Fossils provide a window into the past, offering tangible proof of life forms that lived millions of years ago and illustrating the gradual changes that species underwent over geological time. Transitional forms, in particular, bridge the gaps between major evolutionary stages, shedding light on the evolutionary pathways of various organisms. Here's a detailed analysis of the fossil record and transitional forms:

1. **Formation of Fossils:**

i. **Preservation Conditions:** Fossilization occurs under specific conditions where organisms are rapidly buried, often in sedimentary layers, protecting them from decomposition and external factors.

ii. **Types of Fossils:** Fossils can range from imprints, molds, and casts to petrified remains where organic material is replaced by minerals, preserving the structure of the organism.

2. **Transitional Forms:**

i. **Definition:** Transitional forms, also known as intermediate forms or missing links, exhibit characteristics of both ancestral and descendant groups. They provide crucial evidence of evolutionary relationships.

ii. **Importance:** Transitional forms help bridge gaps in the evolutionary timeline, demonstrating the gradual changes and adaptations that occurred in lineages over time.

3. **Iconic Transitional Fossils:**

i. **Tiktaalik:** Often cited as a significant transitional form, Tiktaalik is a fish-tetrapod intermediate. It lived around 375 million years ago and exhibited features of both fish (such as scales and gills) and tetrapods (limb-like fins).

ii. **Archaeopteryx:** A dinosaur-bird transitional form, lived approximately 150 million years ago. It had feathers and wing-like structures, bridging the gap between dinosaurs and modern birds.

iii. **Ambulocetus:** An ancient whale transitional form, had characteristics of both terrestrial mammals and modern whales. It lived around 50 million years ago and displayed adaptations for both land and aquatic life.

4. **Human Evolutionary Transitions:**

i. ***Australopithecus afarensis*** **(Lucy):** Lucy is a well-known hominid fossil that lived around 3.2 million years ago. With both ape-like and human-like features, Lucy provides insight into early hominid evolution.

ii. ***Homo habilis*:** Dating back to around 2.3 million years ago, is considered one of the earliest members of the Homo genus. It displayed increased brain size and tool use, marking a significant transition in human evolution.

5. **Whale Evolutionary Transitions:**

i. **Pakicetus:** An early whale ancestor, lived approximately 50 million years ago. It had land mammal characteristics but also aquatic adaptations, indicating a transition from terrestrial to aquatic life.

ii. **Basilosaurus:** From around 40 million years ago, was a fully aquatic whale but retained small hind limbs, showcasing vestigial structures from its land-dwelling ancestors.

6. **Critiques and Confirmations:**

i. **Critiques:** Some critics have challenged the concept of transitional forms, often focusing on incomplete fossil records, or suggesting alternative interpretations. However, the discovery of numerous transitional fossils over the years has largely addressed these criticisms.

ii. **Confirmations:** Continued fossil discoveries have not only confirmed the existence of transitional forms but also expanded our understanding of the complexity of evolutionary transitions in various lineages.

7. **Patterns and Evolutionary Trends:**

i. **Adaptive Radiation:** Fossil records show instances of rapid diversification and adaptive radiation, where a single ancestor species gives rise to multiple diverse species. The Galápagos finches are a classic example.

ii. **Mass Extinctions:** Mass extinction events, such as the Permian-Triassic and Cretaceous-Paleogene extinctions, are evident in the fossil record. These events led to significant evolutionary turnover, shaping the course of life on Earth.

8. **Paleogeography and Evolution:**

i. **Continental Drift:** The movement of continents over geological time influenced the distribution of species and led to the formation of distinct ecosystems. Fossil evidence supports the correlation between continental drift and evolutionary divergence.

ii. **Island Biogeography:** Fossil records from islands often exhibit unique evolutionary trajectories due to isolation, leading to the development of endemic species found nowhere else on Earth.

In fact, the fossil record, especially the presence of transitional forms, provides compelling evidence for the evolutionary processes that have shaped life on Earth.

These fossils serve as snapshots of ancient life, demonstrating the intricate pathways through which organisms diversified and adapted to changing environments. The continual discovery of new fossils and the refinement of dating techniques further enhance our understanding of evolutionary patterns and the rich tapestry of life's history.

6.2. COMPARATIVE ANATOMY AND HOMOLOGY

Comparative anatomy and the concept of homology are fundamental aspects of evolutionary biology. They provide crucial evidence for the common ancestry of living organisms and the divergent evolutionary pathways that have led to the incredible diversity of life on Earth. Here's a detailed analysis of comparative anatomy and homology (Figure 6.2):

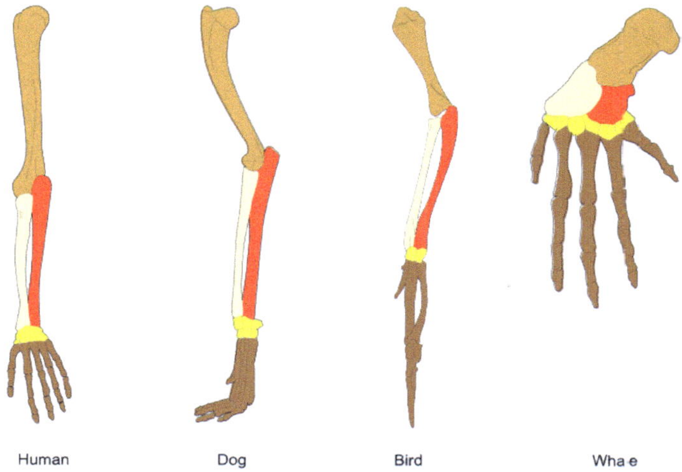

Figure 6.2. Comparative anatomy and homology.

Source: *By Волков Владислав Петрович – <span class="int-own-work" lang="en">Own work</span>, <a href="https://creativecommons.org/licenses/by-sa/4.0" title="Creative Commons Attribution-Share Alike 4.0">CC BY-SA 4.0</a>, <a href="https://commons.wikimedia.org/w/index.php?curid=37704829">Link</a>; Creative Commons License.*

1. **Definition of Comparative Anatomy:** Comparative anatomy involves the study of the similarities and differences in the structures of organisms. By comparing the anatomical features of different species, scientists can infer evolutionary relationships and understand the evolutionary modifications that have occurred over time.
2. **Homologous Structures:**

i. **Definition:** Homologous structures are anatomical features that are similar in structure and origin but may serve different functions in different organisms. These similarities indicate a common evolutionary ancestry.

ii. **Example:** The pentadactyl limb, found in mammals, birds, reptiles, and amphibians, is a classic example of a homologous structure. Despite the different functions (e.g., walking, flying, swimming), the underlying skeletal structure remains similar, suggesting a shared evolutionary origin.

3. **Vestigial Structures:**

i. **Definition:** Vestigial structures are remnants of organs or structures that had a function in the organism's ancestors but have reduced or no function in the modern organism.

ii. **Examples:** Human appendix and tailbone (coccyx) are considered vestigial structures, indicating evolutionary ancestors had functional uses for these structures, while their function in humans has diminished.

4. **Analogous Structures:**

i. **Definition:** Analogous structures are similar in function and appearance but have different evolutionary origins. They arise due to convergent evolution, where different lineages evolve similar traits independently in response to similar environmental challenges.

ii. **Example:** The wings of bats and birds are analogous structures. While both are used for flight, the wing structures and underlying bones are different, reflecting their separate evolutionary paths (Figure 6.3).

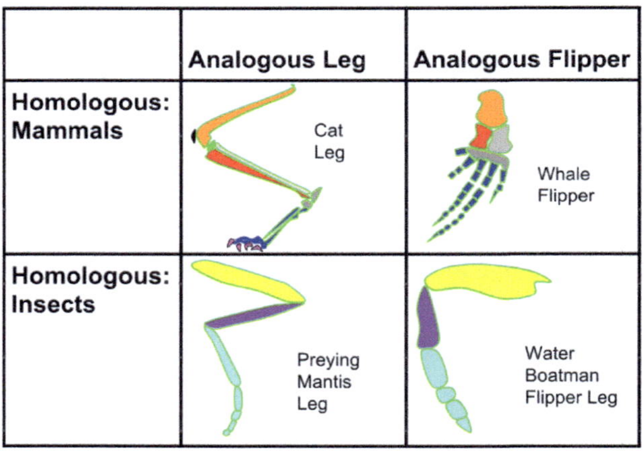

Figure 6.3. Analogous and homologous structures.

Source: *By Vanessablakegraham – Own work, CC BY-SA 4.0, https://commons.wikimedia.org/w/index.php?curid=49147917; Creative Commons License.*

5. **Embryological Homologies:**

i. **Definition:** Embryological homologies refer to similarities in the embryonic development of different species. Shared embryonic structures and processes among organisms reveal their evolutionary relationships.

ii. **Example:** The presence of pharyngeal pouches in the embryonic development of various vertebrates, including fish, amphibians, reptiles, birds, and mammals, is an embryological homology. These structures give rise to different organs (such as gills in fish and tonsils in humans) but originate from a common ancestral structure.

6. **Molecular Homologies:**

i. **Definition:** Molecular homologies involve similarities in DNA, RNA, or protein sequences among different species. Molecular homologies are used to construct phylogenetic trees, illustrating the evolutionary relationships between organisms.

ii. **Example:** Hemoglobin, the oxygen-carrying protein in red blood cells, shows molecular homology across diverse vertebrate species. The amino acid sequences of hemoglobin are more similar among closely related species, reflecting their shared evolutionary history.

7. **Functional Morphology and Evolution:**

i. **Functional Constraints:** Comparative anatomy also considers the functional aspects of biological structures. Evolutionary modifications often occur within the constraints of the organism's existing anatomy and physiology, leading to adaptations that enhance survival and reproduction in specific environments.

ii. **Examples:** The specialized beaks of different bird species, adapted for various feeding habits (e.g., seed-cracking, probing for insects, or catching fish), showcase functional adaptations within the constraints of their common avian ancestry.

8. **Biogeography and Comparative Anatomy:**

i. **Biogeographical Patterns:** Comparative anatomy, when coupled with biogeographical data, provides insights into the geographical distribution of organisms and their evolutionary history. Similar anatomical features among species in specific regions can indicate shared ancestry.

ii. **Example:** Marsupials, such as kangaroos and koalas, are primarily found in Australia. Comparative anatomy, combined with molecular evidence, supports the hypothesis that marsupials share a common ancestry and originated in South America before migrating to Australia.

9. **Implications for Evolutionary Theory:**

i. **Common Ancestry:** Comparative anatomy and the identification of homologous structures provide compelling evidence for the theory of common ancestry, indicating that diverse organisms share a common evolutionary lineage.

ii. **Divergent Evolution:** Homologous structures also demonstrate the process of divergent evolution, where related species evolve different traits from a common ancestor due to environmental pressures and selective forces.

10. **Applications in Evolutionary Research:**

i. **Phylogenetic Studies:** Comparative anatomy is fundamental in constructing phylogenetic trees, depicting the evolutionary relationships between species and higher taxa. Homologous structures provide essential data for these analyses.

ii. **Paleontological Studies:** Fossil comparative anatomy helps paleontologists reconstruct ancient organisms, revealing the evolutionary changes that occurred over millions of years.

Thus, comparative anatomy and the concept of homology provide compelling evidence for evolutionary theory. These fields of study not only illuminate the shared ancestry of diverse life forms but also reveal the remarkable adaptations and modifications that have occurred throughout evolutionary history. The integration of anatomical, embryological, molecular, and functional data continues to enrich our understanding of the evolutionary processes that have shaped the biodiversity of life on Earth.

6.3. MODERN SYNTHESIS

The Modern Synthesis, also known as the Neo-Darwinian Synthesis, refers to the integration of Mendelian genetics with Darwinian evolution that occurred in the early 20th century. This synthesis brought together principles from genetics, natural selection, and population genetics, providing a comprehensive framework for understanding evolutionary processes. Here's a detailed exploration of the Modern Synthesis:

1. Integration of Darwinian Evolution and Mendelian Genetics:

i. **Darwin's Natural Selection:** Charles Darwin's theory of natural selection proposed that species evolve over time through the differential survival and reproduction of individuals with advantageous traits.

ii. **Mendelian Genetics:** Gregor Mendel's laws of inheritance, discovered in the late 19th century, explained how traits are passed from one generation to the next through discrete units (genes).

2. **Key Contributors to the Modern Synthesis:**

i. **Fisher, Haldane, and Wright:** R. A. Fisher, J. B. S. Haldane, and Sewall Wright, among others, laid the mathematical foundations of population genetics, demonstrating how genetic variation within populations could be explained and modeled.

ii. **Dobzhansky and Mayr:** Theodosius Dobzhansky and Ernst Mayr emphasized the importance of natural selection in shaping populations and species. Dobzhansky's work on fruit flies demonstrated evolutionary changes occurring through genetic mutations and natural selection.

3. **Population Genetics and Allele Frequencies:**

i. **Hardy-Weinberg Equilibrium:** The Modern Synthesis incorporated the Hardy-Weinberg equilibrium, a fundamental concept in population genetics. This equilibrium describes the stable frequencies of alleles in an idealized, non-evolving population.

ii. **Genetic Drift and Gene Flow:** The synthesis considered the roles of genetic drift (random changes in allele frequencies) and gene flow (migration of individuals between populations) in shaping genetic diversity.

4. **Mutation and Genetic Variation:**

i. **Mutation as a Source of Variation:** The Modern Synthesis acknowledged mutations as the ultimate source of genetic variation. New alleles arising from mutations provide the raw material upon which natural selection acts.

ii. **Genetic Variation and Adaptation:** Genetic variation within populations allows for adaptation to changing environments. Natural selection acts on advantageous alleles, leading to their increase in frequency over generations.

5. **Speciation and Reproductive Isolation:**

i. **Reproductive Barriers:** The synthesis explained the importance of reproductive isolation mechanisms, such as geographical, behavioral, or genetic barriers, in preventing gene flow between species. These mechanisms lead to the formation of distinct species.

ii. **Sympatric Speciation:** The modern synthesis considered instances of speciation occurring within the same geographic area, challenging earlier ideas that speciation only occurred in geographically isolated populations.

6. **Adaptation and Fitness:**

i. **Adaptation:** The synthesis emphasized that natural selection acts on traits that enhance an organism's fitness, allowing it to survive and reproduce in its specific environment. Adaptations are features that increase an organism's reproductive success.

ii. **Fitness Landscapes:** The concept of fitness landscapes illustrated the relationship between genotypes, phenotypes, and fitness. Populations move through these landscapes, exploring different genetic combinations under the influence of natural selection.

7. **Application to Evolutionary Patterns:**

i. **Gradualism:** The modern synthesis supported the idea of gradual evolutionary change over long periods. Small, incremental changes in allele frequencies within populations lead to macroevolutionary patterns observed in the fossil record.

ii. **Punctuated Equilibrium:** While the modern synthesis emphasized gradual change, it also accommodated the concept of punctuated equilibrium, where long periods of stability are interrupted by brief periods of rapid evolution.

8. **Critiques and Extensions:**

i. **Genetic Drift and Neutral Theory:** Critics highlighted the significance of genetic drift, especially in small populations, challenging the earlier emphasis on natural selection. The neutral theory of molecular evolution proposed that many genetic changes are selectively neutral, having no effect on fitness.

ii. **Extended Evolutionary Synthesis:** Recent developments in fields like epigenetics and developmental biology have led to discussions about extending the Modern Synthesis. The Extended

Evolutionary Synthesis incorporates these additional factors into our understanding of evolutionary processes.

9. **Modern Synthesis and Molecular Biology:**

i. **Genomics and Molecular Evolution:** Advances in genomics and molecular biology have provided detailed insights into genetic mechanisms. Comparative genomics allows researchers to compare entire genomes, revealing evolutionary relationships and conserved genes.

ii. **Evo-Devo:** Evolutionary developmental biology (Evo-Devo) explores how changes in developmental processes contribute to evolutionary innovations. Understanding the regulation of genes during development enhances our understanding of evolutionary change.

So, the Modern Synthesis laid the foundation for modern evolutionary biology, unifying diverse fields of study under a common framework. It provided a comprehensive understanding of how genetic variation, natural selection, and other evolutionary forces shape the diversity of life on Earth. While it has been subject to critiques and extensions, the Modern Synthesis remains a crucial cornerstone of evolutionary theory, guiding research, and exploration into the mechanisms of evolution.

6.4. HISTORY OF MOLECULAR EVOLUTION

Molecular evolution, a branch of evolutionary biology, focuses on the evolutionary changes at the level of genes, proteins, and other molecular structures. It provides insights into the mechanisms governing genetic variation, adaptation, and the relationships between different species. Here's a detailed account of the history of molecular evolution:

1. Early Concepts and Foundations:

i. **Biochemical Comparisons:** In the early 20th century, scientists like Archibald Garrod explored inborn errors of metabolism, discovering that specific diseases were linked to enzyme deficiencies. This hinted at a biochemical basis for heredity and variation.

ii. **Proteins and Amino Acids:** Frederick Sanger's work in the 1950s laid the foundation for understanding the structure of proteins and their constituent amino acids. This knowledge was crucial for later molecular studies.

2. **The Discovery of DNA Structure:**

i. **Watson and Crick:** In 1953, James Watson and Francis Crick elucidated the structure of DNA, showcasing its double-helix form. This discovery provided the molecular basis for inheritance and set the stage for understanding DNA evolution.

3. **The Molecular Clock Hypothesis:**

i. **Zuckerkandl and Pauling:** Emile Zuckerkandl and Linus Pauling proposed the molecular clock hypothesis in the 1960s. This theory suggested that the rate of molecular evolution, particularly amino acid substitutions in proteins, could be used to estimate the divergence times between species.

ii. **Kimura's Neutral Theory:** Motoo Kimura's neutral theory of molecular evolution, proposed in the 1960s, suggested that the majority of molecular changes at the DNA level are selectively neutral and evolve due to genetic drift.

4. **DNA Sequencing and Genetic Codes:**

i. **Sanger Sequencing:** The development of DNA sequencing techniques by Frederick Sanger and colleagues in the 1970s revolutionized molecular biology. It allowed scientists to sequence DNA, uncovering the genetic codes of various organisms.

ii. **Genetic Codes:** The deciphering of genetic codes enabled researchers to translate DNA sequences into amino acids, facilitating comparative analyses of proteins across species.

5. **Molecular Phylogenetics and Systematics:**

i. **Phylogenetic Analysis:** Carl Woese and colleagues pioneered molecular phylogenetics in the late 1970s and early 1980s. They used ribosomal RNA sequences to construct phylogenetic trees, providing a new perspective on the evolutionary relationships between organisms.

ii. **Mitochondrial DNA:** Studies of mitochondrial DNA, which is maternally inherited, provided valuable insights into the evolutionary history of species, especially in animals.

6. Genomic Era and Comparative Genomics:

i. **Genomics and Whole Genome Sequencing:** The Human Genome Project, completed in 2003, marked a significant milestone. Rapid advancements in sequencing technologies led to the sequencing of various genomes, allowing comparative genomic analyses among species.

ii. **Comparative Genomics:** These studies, comparing entire genomes across species, revealed conserved genes, regulatory elements, and evolutionary changes. It provided insights into the genetic basis of adaptation and the evolution of complex traits.

7. **Molecular Evolution and Evolutionary Developmental Biology (Evo-Devo):**

i. **Evo-Devo:** Evolutionary developmental biology, or Evo-Devo, investigates the genetic and molecular mechanisms underlying the development of organisms. Understanding the evolution of developmental pathways sheds light on morphological evolution.

ii. **Gene Regulation and Evolution:** Studies on gene regulation networks have shown how changes in regulatory sequences can lead to morphological diversity. Evolutionary changes in cis-regulatory elements influence the expression of genes during development.

8. **Population Genetics and Molecular Evolution:**

i. **Molecular Population Genetics:** The field of molecular population genetics explores genetic variation within populations. Techniques like DNA polymorphism analysis and coalescent theory provide insights into how natural selection, genetic drift, and migration shape genetic diversity.

ii. **Adaptation and Molecular Signatures:** Detection of molecular signatures of adaptation, such as selective sweeps and positive selection in specific genes, became possible through population genomic studies.

9. **Ancient DNA and Paleogenomics:**

i. **Ancient DNA Studies:** Advances in ancient DNA techniques have allowed scientists to extract and sequence DNA from ancient specimens, providing insights into the genetic makeup of extinct species and human ancestors.

ii. **Paleogenomics:** These studies ancient genomes, revealing evolutionary events, migrations, and adaptations in ancient populations.

10. **Future Directions and Challenges:**

i. **Single-Cell Genomics:** This allows the analysis of individual cells, revealing cellular diversity within organisms. This technique has implications for understanding development, evolution, and disease.

ii. **Systems Biology and Evolution:** Integrating molecular data into systems biology approaches helps understand the complex interactions between genes, proteins, and other molecules. This holistic view enhances our understanding of evolutionary processes at the molecular level.

In summary, the history of molecular evolution reflects a fascinating journey from understanding the basic structure of DNA to unraveling the complexities of genomes and their evolutionary significance. This field continues to evolve, providing profound insights into the mechanisms that underlie the incredible diversity of life on Earth and the processes that have shaped evolutionary history.

6.5. MOLECULAR EVIDENCE

Molecular evidence plays a pivotal role in shaping our understanding of evolutionary relationships, genetic diversity, and the mechanisms driving the evolution of life on Earth.

Through the study of DNA, RNA, proteins, and other molecules, scientists gain invaluable insights into the evolutionary history of organisms and the fundamental processes governing their development. Here's a detailed exploration of molecular evidence in the context of evolutionary biology:

1. **Genetic Material and Molecular Basis of Life:**

i. **DNA and RNA:** Deoxyribonucleic acid (DNA) and ribonucleic acid (RNA) are the genetic materials of life. DNA carries the genetic instructions for the development, functioning, growth, and reproduction of all known living organisms.

ii. **Genetic Code:** The genetic code, consisting of nucleotide triplets (codons), determines the sequence of amino acids in proteins. This code is universal among all living organisms, providing evidence for a common ancestry.

2. **Comparative Genomics and Homologous Genes:**

i. **Homology:** Comparative genomics identifies homologous genes, which are genes in different species that share a common ancestry. Homologous genes provide evidence of evolutionary relationships between organisms.

ii. **Orthologous and Paralogous Genes:** Orthologous genes are homologous genes in different species that diverged from a common ancestor due to speciation. Paralogous genes, on the other hand, arise from gene duplication events within a genome.

3. **Molecular Phylogenetics and Evolutionary Trees:**

i. **Phylogenetic Analysis:** Molecular phylogenetics uses molecular data to construct evolutionary trees, depicting the relationships between species. Techniques like DNA sequencing and sequence alignment aid in phylogenetic reconstructions.

ii. **Cladistics:** This analysis groups organisms based on shared derived characteristics, identified at the molecular level. Cladograms illustrate evolutionary relationships based on these shared traits.

4. **Molecular Clocks and Divergence Time Estimation:**

i. **Molecular Clock Hypothesis:** Molecular clocks use the rate of molecular change to estimate the time since two species shared a common ancestor. This approach helps in dating evolutionary events.

ii. **Mitochondrial DNA Clock:** Mitochondrial DNA is often used for molecular clock studies due to its relatively rapid mutation rate. It provides insights into recent evolutionary events and relationships.

5. **Functional Genomics and Gene Regulation:**

i. **Gene Expression and Regulation:** Molecular evidence reveals how gene expression patterns and regulatory networks have evolved. Changes in regulatory sequences influence the timing and level of gene expression, leading to evolutionary innovations.

ii. **Gene Function Conservation:** Molecular studies identify conserved genes across species, indicating their essential functions. Changes in non-conserved genes often underlie species-specific adaptations.

6. **Genetic Variation and Adaptation:**

i. **Genetic Polymorphism:** Molecular studies reveal genetic polymorphisms, variations in DNA sequences within populations. Polymorphisms are the basis for adaptation, allowing populations to respond to environmental challenges.

ii. **Selective Sweeps:** Molecular evidence identifies regions of the genome where advantageous alleles have rapidly spread through a population due to natural selection, leading to adaptive evolution.

7. **Ancient DNA and Evolutionary History:**

i. **Paleogenomics:** Ancient DNA analysis allows the study of genetic material from ancient specimens. It provides insights into extinct species, evolutionary relationships, and migration patterns of ancient humans and other organisms.

ii. **Molecular Paleontology:** Molecular evidence complements traditional paleontological studies, offering molecular insights into the biology and evolutionary history of ancient organisms.

8. **Comparative Proteomics and Functional Evolution:**

i. **Proteome Comparisons:** Comparative proteomics explores the similarities and differences in protein expression and function across species. It aids in understanding the functional evolution of proteins and biological processes.

ii. **Structural Genomics:** Comparative analysis of protein structures provides insights into the evolution of protein folds and functions. Structural similarities suggest common ancestry and shed light on the molecular basis of biological processes.

9. **Molecular Evidence in Human Evolution:**

i. **Genomic Comparisons:** Comparative genomics between humans and other primates reveal genetic similarities and differences. Molecular evidence supports the evolutionary divergence of human ancestors from common ancestors shared with chimpanzees and other primates.

ii. **Gene Duplication and Brain Evolution:** Molecular studies identify gene duplications related to brain development and function in humans. Changes in these genes have contributed to the evolution of complex cognitive abilities.

10. **Future Directions and Challenges:**

i. **Single-Cell Genomics:** Advancements in single-cell genomics allow the study of gene expression at the level of individual cells. This approach provides insights into cellular diversity, development, and evolutionary innovations.

ii. **Environmental Genomics:** Studying the genomes of organisms in their natural habitats (environmental genomics) provides a holistic understanding of the interactions between organisms, their genes, and their environments.

In fact, molecular evidence stands as a cornerstone of evolutionary biology, enabling scientists to uncover the intricacies of evolutionary processes at the molecular level. By deciphering the genetic codes and unraveling the complexities of genes and proteins, molecular biology continues to deepen our understanding of the evolutionary history of life and the mechanisms that drive its diversity and adaptation. Ongoing advancements in technology and methodology promise further revelations, enriching our knowledge of the molecular basis of evolution.

6.6. EXTENDED EVOLUTIONARY SYNTHESIS

The Extended Evolutionary Synthesis (EES) represents an ongoing scientific discourse that challenges and extends the traditional framework of the Modern Synthesis in evolutionary biology. While the Modern Synthesis, which integrated Darwinian evolution with Mendelian genetics, laid the foundation for understanding evolution, the EES incorporates new insights from various scientific disciplines, including genomics, epigenetics, developmental biology, ecology, and paleontology. Here's a comprehensive exploration of the extended evolutionary synthesis (Figure 6.4):

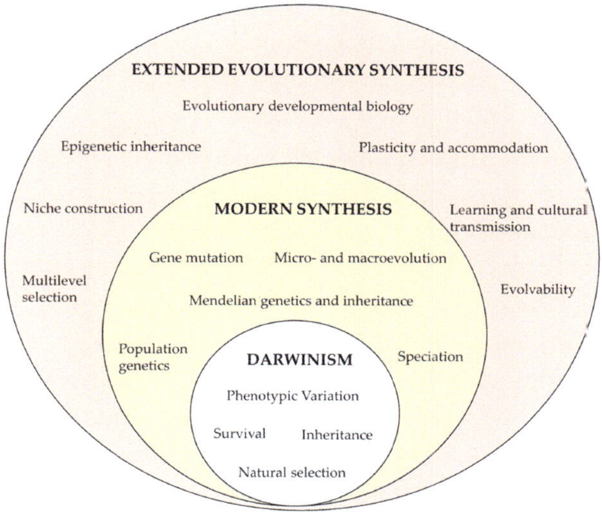

Figure 6.4. Extended evolutionary synthesis.

Source: *Chapelle, V., & Silvestre, F., (2022). Population epigenetics: The extent of DNA methylation variation in wild animal populations. Epigenomes, 6(4), 31. https://doi.org/10.3390/epigenomes6040031; Creative Commons License.*

1. **Critiques of the Modern Synthesis:**

i. **Genetic Determinism:** The modern synthesis often emphasized the role of genes in evolution, potentially oversimplifying the complexity of interactions between genes and the environment.

ii. **Neglect of Non-Genetic Inheritance:** The modern synthesis didn't give substantial attention to non-genetic inheritance mechanisms, such as epigenetic modifications, which can influence evolution.

iii. **Macroevolutionary Processes:** The modern synthesis focused heavily on microevolutionary processes within populations, leaving gaps in our understanding of macroevolutionary patterns and innovations.

2. **Key Concepts of the Extended Evolutionary Synthesis:**

i. **Plasticity and Phenotypic Variation:** EES emphasizes phenotypic plasticity, where organisms can express different traits in response to environmental cues. Plasticity provides a mechanism for organisms to adapt to varying environments.

ii. **Niche Construction:** Organisms not only adapt to their environments but also actively modify and create their environments. Niche construction refers to the process wherein organisms shape their ecological niches, influencing their own evolution and that of other species.

iii. **Epigenetic Inheritance:** EES incorporates the role of epigenetic modifications, heritable changes in gene expression that occur without alterations in the DNA sequence. These modifications can be influenced by environmental factors and affect evolutionary outcomes.

iv. **Evolutionary Developmental Biology (Evo-Devo):** Understanding the genetic and developmental mechanisms that underlie the evolution of phenotypic traits is a central focus of EES. Evo-Devo explores how changes in developmental pathways contribute to evolutionary innovations.

v. **Symbiosis and Multilevel Selection:** EES considers the importance of symbiotic relationships and multilevel selection, where evolution acts not only on individuals but also on groups, populations, and ecosystems.

3. **Genomic Perspectives in EES:**

i. **Comparative Genomics:** This across diverse taxa provides insights into the conservation and divergence of genes and regulatory elements. Understanding conserved genetic elements can reveal shared ancestry, while divergent elements highlight evolutionary innovations.

ii. **Genomic Imprinting:** Where specific genes are expressed based on their parental origin, exemplifies non-Mendelian inheritance. Imprinting can influence the evolution of social behaviors and parental care strategies.

4. **Paleogenomics and Ancient DNA:**

i. **Paleogenomics:** Advances in sequencing ancient DNA allow the study of extinct species and ancient populations. Paleogenomics provides insights into evolutionary processes, migration patterns, and adaptations in ancient organisms.

ii. **Hybridization and Introgression:** Paleogenomic studies reveal instances of hybridization between different species and the exchange of genetic material. Introgression, the transfer of genes between species, plays a significant role in evolutionary dynamics.

5. **Applications and Implications:**

i. **Conservation Biology:** EES has implications for conservation efforts. Recognizing the importance of phenotypic plasticity and ecological interactions helps design conservation strategies that consider the adaptive potential of species facing changing environments.

ii. **Biomedical Research:** Understanding the complexities of genetic and epigenetic interactions informs biomedical research, offering insights into diseases, gene regulation, and personalized medicine.

6. **Challenges and Debates:**

i. **Integration of Disciplines:** One challenge lies in integrating insights from various disciplines within EES. Bridging gaps between genetics, ecology, and development is crucial for a comprehensive understanding of evolution.

ii. **Reconciling with Traditional Views:** EES faces resistance from some scientists who are deeply rooted in the traditional views of the Modern Synthesis. Debates continue on how to best integrate new insights into the existing framework of evolutionary biology.

In summary, the Extended Evolutionary Synthesis represents a paradigm shift in evolutionary biology, incorporating a broader array of mechanisms and disciplines to understand the complexity of evolutionary processes. By embracing the intricacies of genetic and phenotypic interactions, the EES provides a more nuanced and comprehensive view of evolution, enriching our understanding of the diverse and dynamic nature of life's evolutionary journey.

6.7. BIOGEOGRAPHY

Biogeography is the study of the distribution of living organisms and the processes that shape their geographical patterns. It explores the intricate interactions between biological, geological, climatic, and ecological factors, providing valuable insights into the evolution and diversity of life on Earth. Here's a detailed exploration of biogeography and its key aspects (Figure 6.5):

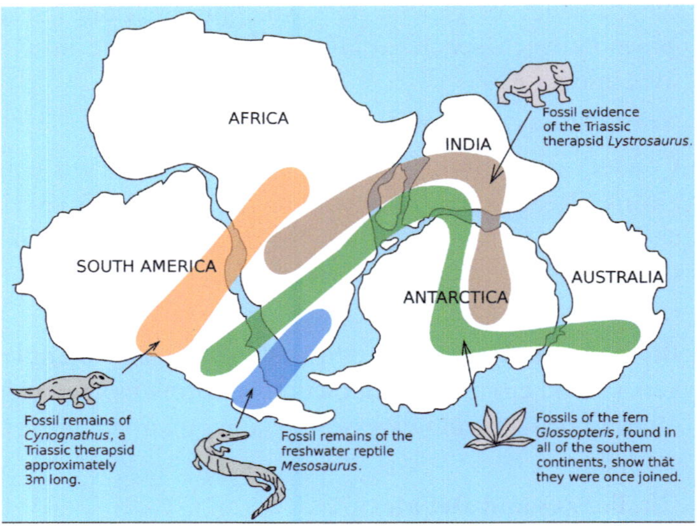

Figure 6.5. Biogeography.

Source: *By Osvaldocangaspadilla – Own work, Public Domain, https://commons.wikimedia.org/w/index.php?curid=11310183; Creative Commons License.*

1. **Historical Development of Biogeography:**

i. **Early Exploration:** Biogeography traces its roots back to the voyages of naturalists like Charles Darwin and Alfred Russel Wallace, who observed and documented the geographical distribution of species.

ii. **Island Biogeography:** Studies of island ecosystems, pioneered by MacArthur and Wilson, provided foundational concepts. The theory of island biogeography explains species richness on islands based on immigration, extinction rates, and island size.

2. **Types of Biogeographical Patterns:**

i. **Disjunct Distribution:** This refers to species occurring in separate, non-contiguous geographic areas. This phenomenon often provides clues about past geological events, climate changes, and continental drift.

ii. **Endemism:** Endemic species are those unique to specific regions and not found anywhere else. Endemism often occurs in isolated ecosystems like islands or remote habitats, highlighting the importance of local adaptation.

3. **Factors Influencing Biogeographical Patterns:**

i. **Plate Tectonics:** The movement of continents and the formation of land bridges and barriers due to plate tectonics have profoundly influenced the distribution of species over millions of years.

ii. **Climate and Ecological Niches:** Climate, including temperature, rainfall, and seasonality, determines the types of ecosystems that can exist in a particular region. Species adapt to specific ecological niches within these climates.

iii. **Dispersal and Migration:** Organisms disperse to new areas, either actively or passively. Dispersal mechanisms, such as wind, water, or animal transport, shape the distribution of species.

iv. **Biological Interactions:** Interactions between species, such as competition, predation, and mutualism, influence their distributions. Ecological factors often lead to the formation of unique communities in different regions.

4. **Biogeographical Realms and Regions:**

i. **Realms:** Biogeographical realms, like the Nearctic, Neotropical, and Palearctic, are large geographic regions characterized by distinct plant and animal assemblages. Each realm represents a unique evolutionary history.

ii. **Biodiversity Hotspots:** Biodiversity hotspots are regions with exceptionally high levels of species richness and endemism. Conservation efforts often target these areas due to their ecological significance.

5. Applications in Conservation and Evolutionary Biology:

i. **Conservation Biogeography:** Biogeographical studies inform conservation strategies by identifying areas of high biodiversity and endemism. Conservation biogeography helps prioritize regions for conservation efforts.

ii. **Phylogeography:** It integrates genetic data with biogeographical patterns, elucidating the evolutionary history and genetic diversity of species within specific geographic areas.

6. Human Impact and Biogeographical Shifts:

i. **Anthropogenic Effects:** Human activities, including habitat destruction, climate change, and the introduction of invasive species, have led to biogeographical shifts. Species are forced to adapt or migrate due to these changes, impacting ecosystems.

ii. **Biogeographical Responses to Climate Change:** As the climate changes, species migrate to maintain suitable habitats. Biogeography helps predict and understand these shifts, aiding in climate change adaptation strategies.

7. **Future Directions and Challenges:**

i. **Global Biogeography:** Advances in satellite technology and big data allow scientists to study biogeographical patterns at a global scale, providing a comprehensive understanding of Earth's biodiversity.

ii. **Integrating Biogeography with Other Disciplines:** Collaborations between biogeographers, geneticists, ecologists, and climatologists enhance our understanding of how biological and environmental factors interact to shape life's distribution.

Thus, biogeography is a multidisciplinary science that illuminates the intricate connections between life and the Earth's dynamic environment. By studying the distribution of species and understanding the factors shaping their geographic patterns, biogeography informs conservation efforts, elucidates evolutionary processes, and deepens our appreciation of the rich tapestry of life across the planet.

6.7.1. Biogeography's Significant Importance in the Study of Evolution

1. **Understanding Diversification:** Biogeographical patterns provide insights into the historical processes of speciation and extinction. By analyzing the distribution of species, scientists can infer past geological events and environmental changes, helping to unravel the evolutionary history of organisms.
2. **Detecting Evolutionary Relationships:** Similar biogeographical patterns among different species often indicate shared evolutionary histories. Studying these patterns aids in constructing phylogenetic trees and understanding the relationships between different taxa, contributing to our knowledge of evolutionary lineages.
3. **Examining Adaptive Radiations:** Biogeography helps explain adaptive radiations, where a single ancestor species rapidly diversified into various forms to exploit different ecological niches. Understanding the geographical context of these radiations provides crucial information about the ecological factors driving evolution.
4. **Revealing Evolutionary Processes:** Biogeography sheds light on evolutionary processes such as migration, dispersal, and vicariance events. These events influence the distribution of genes and traits within populations, leading to genetic differentiation and the formation of new species.
5. **Predicting Biodiversity Patterns:** Biogeographical studies contribute to predicting biodiversity patterns in different regions. Understanding how historical, environmental, and ecological factors shape biodiversity helps in conservation planning and predicting how species might respond to future environmental changes.
6. **Studying Island Biogeography:** Islands often serve as natural laboratories for evolutionary studies due to their unique ecological conditions. The study of island biogeography has provided critical insights into evolutionary processes, including adaptive evolution, speciation, and extinction dynamics.
7. **Explaining Disjunct Distributions:** Disjunct distributions, where related species are found in geographically separated areas, offer clues about continental drift, land bridge formation,

and other geological events. Such patterns help explain how organisms dispersed and diversified over geological time scales.

8. **Examining Evolutionary Responses to Climate Change:** Biogeographical data aids in understanding how species have responded to past climate change events. By studying historical distribution shifts, scientists can infer evolutionary adaptations and migration patterns, providing valuable context for current and future climate change impacts.

Therefore, biogeography serves as a crucial tool in evolutionary studies, enabling scientists to explore the historical, ecological, and environmental factors that have shaped the evolution of life on Earth. Through the analysis of biogeographical patterns, scientists gain valuable insights into the mechanisms driving evolution and the processes that have led to the remarkable biodiversity observed today.

REFERENCES

1. Carl, Z. *The Tangled Bank: An Introduction to Evolution.*
2. Carl, Z., & Douglas, J. E. *Evolution: Making Sense of Life.*
3. Jay, H., Kevin, C., & Zander, C. *Evolution: The Story of Life on Earth.*
4. Jerry, A. C. *Why Evolution is True.*
5. John, M. S., & Eörs, S. *The Origins of Life: From the Birth of Life to the Origin of Language.*
6. Jonathan, W. *The Beak of the Finch: A Story of Evolution in Our Time.*
7. Matt, R. *The Red Queen: Sex and the Evolution of Human Nature.*
8. Neil, S. *Your Inner Fish: A Journey into the 3.5-Billion-Year History of the Human Body.*
9. Paul, G. F. *Life's Engines: How Microbes Made Earth Habitable.*
10. Richard, D. *The Ancestor's Tale: A Pilgrimage to the Dawn of Evolution.*
11. Richard, D. *The Blind Watchmaker.*
12. Richard, D. *The Greatest Show on Earth.*
13. Sean, B. C. *Endless Forms Most Beautiful: The New Science of Evo Devo.*
14. Sean, B. C. *Remarkable Creatures: Epic Adventures in the Search for the Origins of Species.*
15. Sean, B. C. *The Making of the Fittest: DNA and the Ultimate Forensic Record of Evolution.*

CHAPTER 7

Philosophy of Evolution

CONTENTS

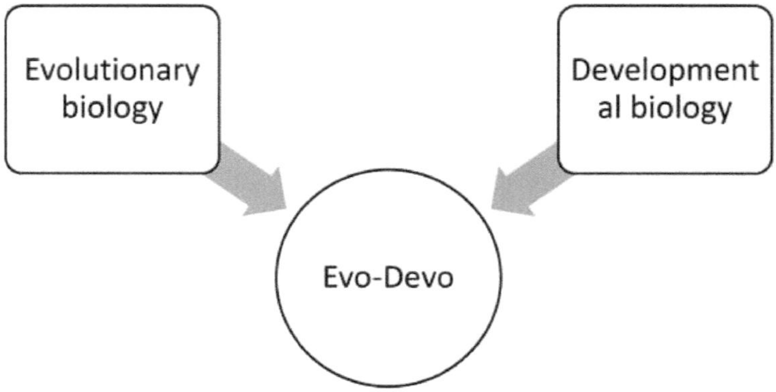

Figure 7.1. Philosophy of evolution.

Source: Marcum, J. A., (2017). Evolutionary philosophy of science: A new image of science and stance towards general philosophy of science. Philosophies, 2(4), 25. https://doi.org/10.3390/philosophies2040025; Creative Commons License.

7.1. DARWINISM

Darwinism, often used interchangeably with the theory of evolution, specifically refers to the evolutionary theory proposed by Charles Darwin. It encompasses a broad range of ideas and concepts, providing a comprehensive understanding of the mechanisms of evolution and their impact on the diversity of life. Here's an in-depth exploration of Darwinism:

1. Natural Selection:
   - **Principle of Natural Selection:** Darwinism centers on the principle of natural selection, where organisms with advantageous traits for survival and reproduction in a given environment are more likely to pass these traits to the next generation.
   - **Variation and Differential Reproduction:** Darwin emphasized the importance of heritable variations within populations. Individuals with variations that confer reproductive advantages contribute more offspring to the next generation, leading to the gradual evolution of species.
2. Descent with Modification:
   - **Common Descent:** Darwin proposed that all living organisms share a common ancestry. Over time, as populations accumulate

small variations through natural selection, diverse species emerge, forming a branching tree of life.

- **Adaptation:** Darwinism explains how organisms adapt to their environments over generations. Through natural selection, advantageous traits become prevalent, allowing species to fit into specific ecological niches.

3. Evidence for Evolution:

- **Fossil Record:** Darwinism aligns with the fossil record, demonstrating a chronological sequence of organisms, showcasing transitions between different species over geological time.
- **Comparative Anatomy:** Comparative anatomical structures in different species, such as homologous and vestigial organs, provide evidence of common ancestry, supporting Darwin's ideas.
- **Biogeography:** The geographical distribution of species and the presence of similar organisms in diverse habitats reflect evolutionary processes, consistent with Darwin's predictions.

4. Sexual Selection:

- **Sexual Dimorphism:** Darwinism introduced the concept of sexual selection, where traits that enhance an individual's mating success, even at the expense of survival, can evolve. This explains the evolution of elaborate traits in certain species.
- **Mate Choice:** Darwin proposed that individuals often choose mates based on specific traits. This selection can lead to the evolution of traits that confer advantages in reproduction.

5. Modern Synthesis and Genetic Basis:

- **Integration with Genetics:** The modern synthesis combined Darwinian evolution with Mendelian genetics, explaining how genetic variations contribute to evolutionary changes. Mutations, genetic drift, and recombination are integrated into the framework of Darwinism.
- **Population Genetics:** Darwinism incorporates principles from population genetics, emphasizing the role of genetic variations within populations and how they are subject to natural selection.

6. Continued Developments and Debates:

- **Extended Evolutionary Synthesis:** Darwinism continues to evolve with the Extended Evolutionary Synthesis, integrating new findings from genomics, epigenetics, and developmental biology. This expansion enriches our understanding of evolutionary processes.

- **Debates and Challenges:** Darwinism faces debates, including discussions about the rate of evolution, punctuated equilibrium, and the role of symbiosis. These debates contribute to the ongoing refinement of evolutionary theory.

In summary, Darwinism represents a foundational theory in biology, providing a robust framework to understand the processes and patterns of evolution. It has stood the test of time and remains a cornerstone of modern evolutionary biology, shaping our understanding of the unity and diversity of life on Earth.

7.2. ALTERNATIVES

7.2.1. Catastrophism

Catastrophism is a scientific theory that suggests that the Earth has been primarily shaped by sudden, short-lived, and violent events, often referred to as catastrophes. These catastrophic events are believed to have caused significant geological and biological changes in a relatively short period of time. Here's a detailed exploration of catastrophism:

1. Historical Context:

- **Early Notions:** Catastrophism has ancient roots, with early civilizations attributing geological features to divine or catastrophic events.
- **Revival in the 19th Century:** Catastrophism gained prominence in the 19th century, partly as a response to the emerging theory of uniformitarianism, which suggested slow and gradual geological processes.

2. Key Principles of Catastrophism:

- **Sudden Events:** Catastrophism posits that major geological and biological changes are primarily the result of sudden and intense events, such as floods, volcanic eruptions, asteroid impacts, or earthquakes.
- **Limited Duration:** Catastrophes are thought to be relatively

short-lived events, leading to rapid and dramatic alterations in the Earth's landscape and ecosystems.

- **Environmental Impact:** Catastrophes are believed to have a profound impact on the environment, causing mass extinctions, the formation of new landforms, and significant climatic changes.

3. Catastrophism and Paleontology:

- **Mass Extinctions:** Catastrophism is often invoked to explain mass extinctions in the fossil record, such as the Cretaceous-Paleogene extinction event that led to the demise of dinosaurs.
- **Rapid Fossilization:** Catastrophic events, like volcanic eruptions or tsunamis, are proposed as explanations for rapid fossilization processes, preserving organisms in unusual conditions.

4. Catastrophism and Geology:

- **Formation of Geological Features:** Catastrophism is used to explain the rapid formation of geological features, such as mountains, canyons, and sedimentary layers, as a result of sudden tectonic shifts or catastrophic flooding.
- **Deposition of Sediments:** Catastrophes, like massive floods, are believed to explain the rapid deposition of sediments and the formation of geological strata.

5. Challenges and Modern Interpretations:

- **Uniformitarianism vs. Catastrophism:** The debate between uniformitarianism (the idea that geological processes occur gradually over long periods) and catastrophism continues. Modern geology often integrates elements of both theories, recognizing the significance of both gradual and sudden events.
- **Impact of New Discoveries:** Discoveries related to meteorite impacts, ancient tsunamis, and climate change have prompted a reevaluation of catastrophism, leading to a more nuanced understanding of how sudden events have shaped the Earth's history.

6. Catastrophism in Modern Science:

- **Planetary Science:** Catastrophism is studied in the context of other celestial bodies, such as the Moon and Mars, where impacts and volcanic events have shaped their surfaces.
- **Climate Change and Environmental Catastrophes:**

Catastrophism is invoked in discussions about the potential environmental catastrophes caused by human-induced climate change, emphasizing the rapid and widespread impacts on ecosystems.

In summary, while the strict interpretation of catastrophism as the sole driver of geological and biological change has been largely replaced by gradualist and uniformitarian explanations, the concept of catastrophism continues to play a role in explaining significant geological and biological events throughout Earth's history. Modern science recognizes the interplay of gradual processes and sudden events, leading to a more nuanced understanding of the Earth's dynamic and complex history.

7.2.2. Lamarckism

Lamarckism, named after the French biologist Jean-Baptiste Lamarck, is an early theory of evolution that posits that acquired traits during an organism's lifetime can be passed on to its offspring. This idea, often summarized as the theory of inheritance of acquired characteristics, has historical significance in the development of evolutionary thought. Here's a detailed exploration of Lamarckism:

1. Inheritance of Acquired Characteristics:

- **Core Principle:** Lamarckism proposes that organisms can acquire new traits or modify existing ones in response to their environment or needs during their lifetime. These acquired traits are then passed on to the next generation.
- **Example:** Lamarck used the elongated neck of giraffes as an example. He suggested that giraffes developed long necks over generations because they continually stretched their necks to reach high branches. According to Lamarckism, this acquired trait would be inherited by their offspring.

2. Use and Disuse Theory:

- **Useful Traits:** Lamarck argued that traits used frequently by an organism become more developed, while unused traits deteriorate over generations. This principle underlines the adaptability of organisms to their environments.
- **Example:** Lamarck used the example of cave-dwelling animals losing their eyesight over generations due to disuse, suggesting

that these traits would eventually disappear in the descendants of these animals.

3. Complexifying Forces:
    - **Internal Vital Forces:** Lamarck proposed the existence of internal vital forces driving organisms towards complexity and perfection. These forces were believed to guide evolutionary changes in a purposeful manner.
    - **Adaptation to Environment:** Lamarckism emphasized the adaptability of organisms to their environment, suggesting that evolutionary changes occurred in response to the specific needs of the organism in its surroundings.
4. Challenges and Relevance:
    - **Scientific Criticisms:** Lamarckism faced scientific criticism, especially after the rise of Mendelian genetics and the discovery of the role of DNA in inheritance. The theory lacked a mechanism for the inheritance of acquired traits.
    - **Modern Understanding:** While Lamarckism is not considered a valid explanation for evolution, some elements of the theory have found relevance in the study of epigenetics, where environmental factors can influence gene expression without altering the underlying DNA sequence.
5. Lamarckism's Impact:
    - **Historical Significance:** Lamarckism played a crucial role in the early development of evolutionary thought. It represented an attempt to explain the diversity of life before the mechanisms of natural selection and genetics were fully understood.
    - **Evolutionary Theory Evolution:** Lamarckism, despite being outdated as a complete theory, contributed to the evolution of evolutionary theory itself. It paved the way for discussions about adaptation, the role of the environment, and the dynamic nature of living organisms.

In summary, while Lamarckism has been largely supplanted by the modern understanding of evolution through natural selection and genetics, its historical significance lies in its role as an early attempt to explain the mechanism of evolution. Although Lamarckism is not considered a valid scientific theory today, its influence on the development of evolutionary thought continues to be acknowledged in the context of the history of biology.

7.2.3. Orthogenesis

Orthogenesis, also known as orthogenetic evolution or straight-line evolution, is a controversial and largely discredited hypothesis in evolutionary biology. This theory posits that evolution follows a predetermined, straight-line trajectory, leading to progressive, irreversible changes in organisms over time. Here's a detailed exploration of orthogenesis and its historical context (Figure 7.2):

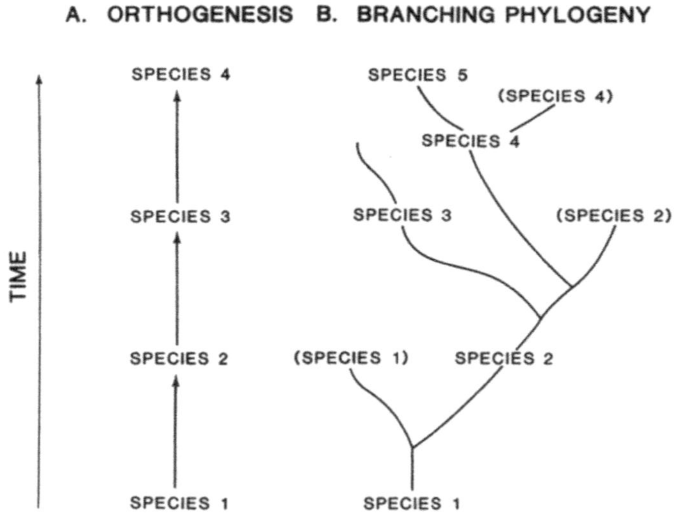

Figure 7.2. Orthogenesis.

Source: MacFadden, B. J., Oviedo, L. H., Seymour, G. M., et al., (2012). Fossil horses, orthogenesis, and communicating evolution in museums. Evo. Edu. Outreach, 5, 29–37. https://doi.org/10.1007/s12052-012-0394-1; Creative Commons License.

1. Historical Roots: Late 19th and Early 20th Century: Orthogenesis gained popularity during the late 19th and early 20th centuries, a time when the mechanisms of evolution were not fully understood. Scientists of the era attempted to explain the complexity and diversity of life through various theories, including orthogenesis.

2. Key Concepts of Orthogenesis:

- **Irreversible Change:** Orthogenesis proposed that evolutionary changes occur in a predetermined, irreversible manner. Species were believed to follow a fixed path, leading to continuous improvement or refinement over generations.

- **Absence of Selection:** Unlike natural selection, orthogenesis did not emphasize the role of external factors or environmental pressures in driving evolutionary changes. Instead, it suggested that intrinsic, internal forces guided species along a specific evolutionary trajectory.

3. Variants of Orthogenesis:

- **Paleontological Orthogenesis:** Some paleontologists proposed that certain lineages exhibited a directional trend in morphology, suggesting a form of orthogenesis. These claims were often based on the perceived progressive changes in fossil records.
- **Psychological Orthogenesis:** In the realm of human evolution, orthogenesis theories were proposed to explain the supposed progression from primitive to advanced mental and psychological traits. These theories have been heavily criticized and rejected by modern anthropology and psychology.

4. Challenges and Criticisms:

- **Lack of Mechanism:** One of the primary criticisms of orthogenesis was the absence of a clear mechanism driving the supposed straight-line evolution. Unlike natural selection, which had a well-defined mechanism (differential reproductive success), orthogenesis lacked explanatory power.
- **Incompatibility with Genetics:** The discovery of Mendelian genetics and the understanding of genetic variation within populations contradicted the notion of predetermined, fixed evolutionary paths. The principles of genetics provided a mechanism for the variability observed in natural populations.

5. Rejection and Modern Perspectives:

- **Decline of Orthogenesis:** By the mid-20th century, orthogenesis lost favor among scientists due to the overwhelming evidence supporting natural selection and genetic variation as the driving forces of evolution.
- **Contemporary Understanding:** Modern evolutionary biology, grounded in the principles of population genetics and natural selection, rejects orthogenesis. Evolutionary changes are

understood as a result of genetic mutations, recombination, and environmental pressures acting on populations.

6. Legacy and Historical Context:

- **Historical Significance:** Orthogenesis, despite its rejection, remains significant in the history of evolutionary thought. It represents a stage in the development of evolutionary theories before the synthesis of Darwinian natural selection and Mendelian genetics.
- **Cautionary Tale:** Orthogenesis serves as a cautionary tale in scientific inquiry, highlighting the importance of empirical evidence and the need for testable mechanisms in the formulation of scientific theories.

In short, orthogenesis, once a prominent theory in evolutionary biology, has been supplanted by the modern synthesis, which integrates natural selection and genetics. While it played a role in the historical development of evolutionary thought, orthogenesis serves as a reminder of the importance of empirical validation and the continuous refinement of scientific theories based on robust evidence and experimental observation.

7.2.4. Mutationism

Mutationism is a historical theory of evolution that suggests that new species arise primarily through sudden, significant mutations. Unlike the gradual changes proposed by Darwin's natural selection, mutationism posits that evolution occurs in large steps due to mutations of considerable effect. Here's a detailed exploration of mutationism and its historical context:

1. Mutation as the Driving Force:

- **Central Idea:** Mutationism proposes that most evolutionary changes happen as a result of large, abrupt, and heritable changes in the genetic material of organisms. These mutations, rather than gradual variations, drive the formation of new species.
- **Role of Selection:** Mutationism acknowledges the role of natural selection but suggests that selection acts mainly to stabilize new mutations once they have arisen, rather than being the primary mechanism for generating variation.

2. Historical Context: Late 19th and early 20th century: Mutationism gained prominence during the late 19th and early 20th centuries, a period marked by significant advancements in genetics. Scientists like Hugo de Vries and

William Bateson supported mutationism, believing that discrete mutations were the primary source of evolutionary change.

3. Key Concepts of Mutationism:

- **Saltation:** Mutationism is associated with the concept of saltation, which refers to significant, discontinuous changes in an organism's phenotype due to mutations. These large leaps were believed to lead to the formation of new species in a single generation.
- **Hopeful Monsters:** A controversial aspect of mutationism was the idea of "hopeful monsters," suggesting that major evolutionary changes could occur in a single generation, leading to the sudden appearance of drastically different organisms.

4. Challenges and Decline:

- **Lack of Evidence:** Mutationism faced criticism due to the lack of observable evidence supporting the occurrence of large, beneficial mutations that could drive substantial evolutionary changes.
- **Integration with Mendelian Genetics:** The integration of mutationism with Mendelian genetics led to the understanding that most mutations were either harmful or neutral, making it challenging for mutationism to explain the gradual, adaptive changes observed in natural populations.

5. Modern Perspectives:

- **Synthesis with Modern Genetics:** Modern evolutionary biology integrates the concept of mutations into the understanding of genetic variation. While mutations can be significant sources of genetic diversity, they are usually small in effect and are subject to the process of natural selection over generations.
- **Punctuated Equilibrium:** The concept of punctuated equilibrium, proposed by Stephen Jay Gould and Niles Eldredge, suggests that evolutionary change occurs in relatively rapid bursts (punctuations) separated by long periods of stability. While not mutationism per se, it shares the idea of rapid evolutionary change.

6. Legacy and Historical Significance:

- **Historical Influence:** Despite its decline as a standalone theory, mutationism contributed to the broader understanding of the role of genetic mutations in evolution. It served as a transitional phase between early evolutionary theories and the modern synthesis of the mid-20th century.

- **Scientific Caution:** Mutationism serves as a reminder of the importance of empirical evidence and the need for a comprehensive understanding of the mechanisms of evolutionary change. It highlights the complexity of evolutionary processes and the caution required when proposing radical theories without substantial supporting evidence.

In summary, mutationism was a theory of evolution that emphasized the significance of large, abrupt mutations in driving evolutionary change. While largely supplanted by the modern synthesis, which integrates natural selection and genetics, mutationism played a role in shaping the historical development of evolutionary thought and contributed to the ongoing exploration of genetic mechanisms in evolution.

7.2.5. Saltationism

Saltationism, also known as saltational evolution or leaps and jumps theory, is a historical hypothesis in evolutionary biology that suggests evolutionary changes occur in large, sudden leaps or jumps, rather than through gradual, incremental changes. Here's a detailed exploration of saltationism and its historical context (Figure 7.3):

1. Sudden and Significant Changes:
    - **Core Idea:** Saltationism proposes that evolutionary transformations happen suddenly and in large steps, leading to the abrupt appearance of new species or traits in a single generation.
    - **Role of Mutations:** Saltationism often attributes these large leaps to significant mutations, which can result in the rapid emergence of new characteristics or species without the need for the gradual accumulation of small variations.
2. Historical Background:
    - **Late 19th and Early 20th Century:** Saltationism gained attention during the late 19th and early 20th centuries, partly as a response to the challenges faced by gradualistic explanations of evolutionary change. Scientists like Hugo de Vries and Richard Goldschmidt were prominent advocates of saltationist ideas.
    - **Hopeful Monsters:** A controversial concept associated with saltationism is that of "hopeful monsters," suggesting that drastic

and advantageous changes might result in new, viable species in a single generation, effectively creating a new species in a single leap.

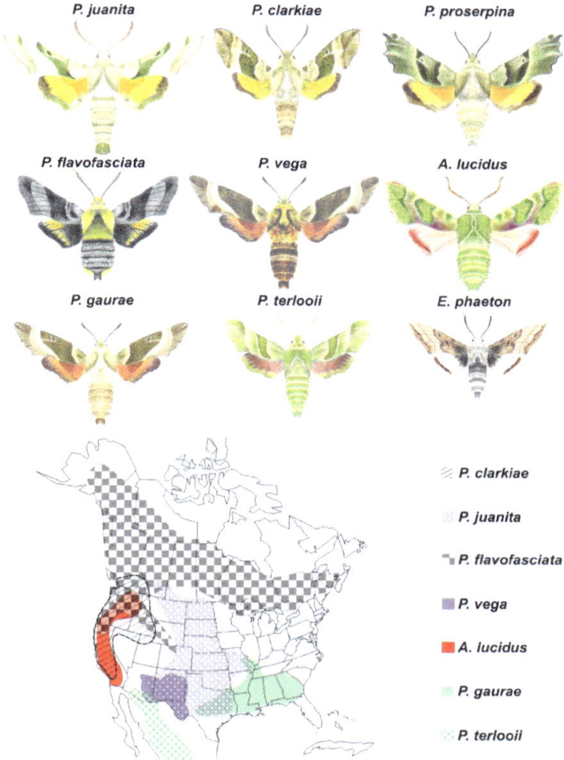

Figure 7.3. Saltationism-as seen in moths.

Source: Rubinoff, D., & Le Roux, J. J., (2008). Evidence of repeated and independent saltational evolution in a peculiar genus of sphinx moths (Proserpinus: Sphingidae). PLoS One, 3(12), e4035. https://doi.org/10.1371/journal.pone.0004035; Creative Commons License.

3. Key Concepts of Saltationism:

- **Macro-Mutations:** Saltationism posits the existence of macro-mutations, which are large-scale genetic changes that occur in a single step and can lead to significant alterations in an organism's phenotype.
- **Rapid Speciation:** Saltationism suggests that new species can arise suddenly and dramatically through these large mutations, challenging the gradualist view of speciation.

4. Challenges and Criticisms:

- **Lack of Evidence:** One of the primary criticisms of saltationism is the lack of empirical evidence supporting the occurrence of large, beneficial mutations that can lead to the formation of new species in a single generation.
- **Genetic Constraints:** The understanding of genetics and the presence of genetic constraints on phenotypic changes made it challenging for saltationism to explain how large mutations could consistently result in functional and adaptive traits.

5. Modern Perspectives:

- **Integration with Modern Understanding:** While saltationism as a standalone theory has been largely discredited, modern evolutionary biology recognizes the occasional occurrence of large-effect mutations. However, these mutations are typically rare and often have deleterious effects, making them unlikely mechanisms for the rapid formation of new, viable species.
- **Punctuated Equilibrium:** The concept of punctuated equilibrium, proposed by Stephen Jay Gould and Niles Eldredge, shares some similarities with saltationism. Punctuated equilibrium suggests that evolutionary change occurs in relatively rapid bursts (punctuations) separated by long periods of stability.

6. Legacy and Historical Significance:

- **Historical Context:** Saltationism represents a stage in the historical development of evolutionary thought, serving as a counterpoint to gradualism and playing a role in the broader discourse on the mechanisms of evolution.
- **Scientific Caution:** Saltationism serves as a reminder of the need for empirical evidence and the importance of understanding the genetic mechanisms underlying evolutionary change. It highlights the complexity of the processes governing the emergence of new traits and species (Figure 7.4).

In summary, saltationism was a historical hypothesis proposing that evolutionary changes occur suddenly and in large steps, often attributed to significant mutations. While it played a role in the historical development of evolutionary thought, saltationism has been largely supplanted by the modern understanding of evolution, which emphasizes the role of gradual, incremental changes and natural selection in the formation of new species and traits.

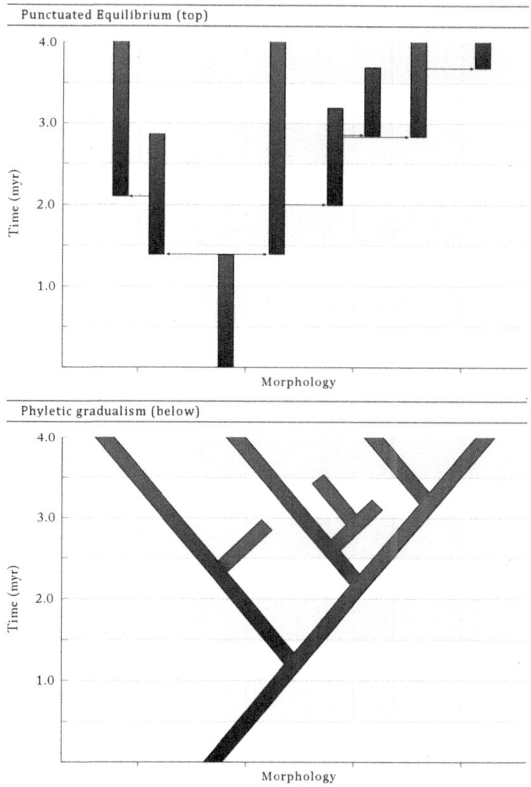

Figure 7.4. Punctuated equilibrium.

Source: By Miguel Chavez – Own work, CC BY-SA 4.0, https://commons.wikimedia.org/w/index.php?curid=65759614; Creative Commons License.

7.2.6. Structuralism (Spandrel)

Structuralism, within the context of evolutionary biology and the concept of spandrels, addresses the idea that some traits or characteristics in organisms might not have originated due to direct adaptation but rather as by-products of other evolutionary changes. This theory, popularized by Stephen Jay Gould and Richard Lewontin, suggests that not all features have adaptive significance and that some traits emerge as incidental consequences of evolutionary processes. Here's a detailed exploration of structuralism and the spandrel concept (Figure 7.5):

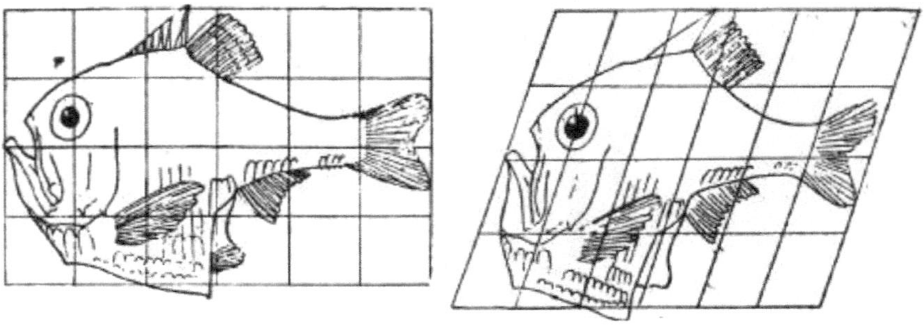

Figure 7.5. Structuralism (spandrel theory).

Source: By D'Arcy Wentworth Thompson – This is an illustration from page 1062, volume II, chapter XVII, of On Growth and Form by D'arcy Wentworth Thompson., Public Domain, https://commons.wikimedia.org/w/index.php?curid=75538458; Creative Commons License.

1. Spandrels and Architectural Analogy:
    - **Spandrels in Architecture:** The term "spandrel" refers to the triangular spaces formed between the curving structures of arches in architecture. Gould and Lewontin used this term metaphorically to describe features in organisms that arise as by-products of other adaptations, without direct adaptive significance.
    - **Exaptation:** Spandrels are often associated with exaptation, where a trait that initially had no adaptive function is co-opted for a new purpose. This concept emphasizes the non-adaptive origins of certain features.
2. Historical Context:
    - **Critique of Adaptationism:** Structuralism emerged as a critique of adaptationism, the idea that every trait in an organism is an adaptation shaped by natural selection. Gould and Lewontin argued that this adaptationist view oversimplified the complexity of evolutionary processes.
3. Key Concepts of Spandrel Theory:
    - **Non-Adaptive Traits:** The spandrel concept posits that some traits might be side effects of other adaptive changes or might be neutral in terms of selection. These traits are not necessarily adaptations in themselves but are products of the evolutionary process.

- **Constraints and Developmental Pathways:** Spandrels highlight the constraints imposed by developmental and genetic processes, which can lead to the emergence of certain traits even without adaptive selection pressure.

4. Examples and Criticisms:

- **Examples of Spandrels:** Gould and Lewontin used various examples, such as the human chin and certain aspects of brain architecture, to illustrate spandrels – traits that might not have direct adaptive significance but are results of other evolutionary changes.
- **Criticism and Debate:** Spandrel theory has been both praised for its nuanced view of evolution and criticized for its potential overemphasis on non-adaptive traits. Some argue that many traits previously considered spandrels might indeed have adaptive significance.

5. Modern Perspectives:

- **Integration with Evolutionary Biology:** Modern evolutionary biology integrates the spandrel concept with the understanding of adaptation and exaptation. Researchers acknowledge the complexity of traits and consider both their origins and potential functions when studying evolutionary biology.
- **Evolutionary Developmental Biology (Evo-Devo):** Spandrel theory aligns with the field of evolutionary developmental biology, which explores how developmental processes influence the evolution of traits. Evo-Devo emphasizes the role of genetic and developmental constraints in shaping phenotypic variation.

6. Legacy and Impact:

- **Scientific Caution:** The spandrel concept serves as a reminder of the need for caution when assuming adaptive significance for every trait. It encourages scientists to consider a broader range of factors, including developmental and historical constraints, when studying the evolution of traits.
- **Continued Exploration:** Spandrel theory has led to continued exploration of the origins of traits, emphasizing the multifaceted nature of evolutionary processes. It has influenced discussions on adaptation, neutral evolution, and the complexity of biological traits.

In summary, structuralism, as exemplified by the spandrel concept, challenges the simplistic view of adaptationism by highlighting the non-adaptive origins of certain traits. While it has faced criticism and sparked debates, the spandrel concept has left a lasting impact on evolutionary biology, encouraging scientists to consider the intricate interplay of developmental, genetic, and adaptive factors in the evolution of biological traits.

7.2.7. Theistic Evolution

Theistic evolution, also known as evolutionary creationism, is the belief that the processes of evolution as described by modern science are compatible with the existence and involvement of a divine Creator. This perspective reconciles the scientific understanding of evolution with religious faith, suggesting that evolution is a tool used by a higher power to bring about the diversity of life on Earth. Here's a detailed exploration of theistic evolution (Figure 7.6):

Figure 7.6. Theistic evolution or evolutionary creationism.

Source: By John Martin – original URL archive URL, Public Domain, https://commons.wikimedia.org/w/index.php?curid=13125299; Creative Commons License.

1. Harmony of Science and Faith:

- **Divine Guidance:** Theistic evolution proposes that God, or a divine force, guided the evolutionary processes over millions of years. Evolution, in this view, is the means through which God created and continues to create the vast array of living beings.

- **Acceptance of Scientific Evidence:** Theistic evolutionists fully accept the scientific consensus on the age of the Earth, common ancestry, and the mechanisms of evolution, including natural selection and genetic variation.

2. Reconciliation with Religious Texts:

- **Symbolic Interpretation:** Theistic evolution often involves a symbolic interpretation of religious texts. Creation stories are viewed as metaphorical or symbolic, conveying spiritual truths rather than literal, scientific accounts of the Earth's origins.
- **God's Immanence:** Theistic evolution emphasizes the immanence of God in the natural world, suggesting that God works through natural laws and processes, including evolution, to accomplish divine purposes.

3. Human Evolution:

- **Special Creation of Humans:** Theistic evolution allows for the possibility that humans, while evolving from earlier species, were specially created by God at some point in the evolutionary process. This perspective addresses theological concerns about the uniqueness and moral responsibility of human beings.

4. Critiques and Controversies:

- **Theological Concerns:** Some religious groups and individuals criticize theistic evolution, arguing that it challenges the concept of a literal Adam and Eve, a central figure in many religious traditions.
- **Scientific Acceptance:** Theistic evolution is sometimes criticized within the scientific community for introducing supernatural elements into the naturalistic framework of evolutionary theory.

5. Integration with Environmental Stewardship:

- **Environmental Ethics:** Theistic evolution often emphasizes the concept of stewardship, wherein humans are seen as caretakers of the Earth and its biodiversity. This perspective encourages environmental responsibility and conservation efforts.
- **Caring for Creation:** Theistic evolutionists argue for the moral responsibility to care for the natural world, viewing it as a divine creation that humans are entrusted to preserve and protect.

6. Diverse Religious Perspectives:

- **Christianity:** Theistic evolution is most commonly associated with certain branches of Christianity, where theologians and religious leaders have advocated for its compatibility with Christian doctrine.
- **Other Religions:** Similar reconciliatory perspectives exist in other religions, such as Reform Judaism and certain interpretations within Islam, where believers see no inherent conflict between evolutionary science and religious faith.

7. Legacy and Impact:

- **Interfaith Dialogue:** Theistic evolution has contributed to interfaith dialogue and discussions between religious communities and the scientific community, fostering understanding and cooperation on issues related to science and faith.
- **Education and Curriculum:** In some religious educational institutions, theistic evolution is included in the curriculum, providing students with an understanding of both scientific principles and religious beliefs.

In fact, theistic evolution represents an attempt to reconcile religious faith with the scientific understanding of evolution. By embracing the scientific consensus while asserting the presence of a divine Creator, theistic evolution seeks to harmonize faith and reason, fostering a nuanced perspective on the origins and development of life on Earth.

7.2.8. Vitalism

Vitalism is a philosophical and scientific concept that posits the existence of a vital force or life essence distinct from physical and chemical processes. Historically, vitalism was an influential idea in the study of biology, suggesting that living organisms were fundamentally different from non-living matter due to the presence of this vital force. Here's a detailed exploration of vitalism and its historical context:

1. Life Force and Vital Essence:

- **Core Idea:** Vitalism proposes the existence of a unique life force or vital essence that distinguishes living organisms from inanimate matter. This force was thought to be responsible for life's processes and organization.

- **Non-Reductionist Perspective:** Vitalism rejects reductionism, the idea that all biological phenomena can be explained by the principles of physics and chemistry alone. Vitalists argue that life cannot be reduced to mere physical and chemical processes.

2. Historical Significance:

- **Early Beliefs:** Vitalism has deep historical roots and was prevalent in ancient and medieval thought. Many cultures had concepts akin to vitalism, attributing life to some form of animating force.
- **Scientific Influence:** Vitalism influenced early biological theories and was considered a viable explanation for life processes before the advent of modern biochemistry and cell biology.

3. Challenges and Criticisms:

- **Lack of Empirical Evidence:** One of the main criticisms of vitalism was the absence of empirical evidence supporting the existence of a life force. Vitalism lacked a scientific basis and was often seen as a mystical or metaphysical concept.
- **Advancements in Biochemistry:** With the discoveries in biochemistry, especially in the 19th and 20th centuries, the chemical basis of life processes was elucidated. Enzymes, proteins, and nucleic acids were found to be essential components of biological functions, rendering vitalism increasingly obsolete.

4. End of Vitalism in Science:

- **Mechanistic Biology:** The rise of mechanistic biology, which explained biological phenomena through the principles of chemistry and physics, led to the decline of vitalism in scientific circles. Biological processes were increasingly understood in terms of chemical reactions and molecular interactions.
- **Modern Biophysics and Molecular Biology:** Advances in biophysics and molecular biology provided detailed insights into the biochemical and physical mechanisms underlying life processes, further diminishing the need for a vital force explanation.

5. Legacy and Philosophical Significance:

- **Philosophical Debates:** Vitalism continues to be a topic of philosophical debate, especially in discussions about the nature of life and consciousness. Some philosophical perspectives explore the idea of emergent properties and holistic views of life that go beyond reductionist explanations.

- **Holistic Medicine and Alternative Therapies:** Some holistic and alternative medicine practices incorporate vitalistic concepts, although these ideas often diverge from mainstream scientific understanding.

In summary, vitalism was a historically significant concept that attempted to explain the unique properties of living organisms through the existence of a vital force. While it influenced early biological thought, vitalism gradually lost credibility in the face of scientific advancements, particularly in biochemistry and molecular biology. While the concept persists in philosophical discussions, the scientific community generally regards vitalism as an outdated and unsupported explanation for the nature of life.

7.2.9. Teleology in Biology

Teleology in biology refers to the interpretation of biological phenomena in terms of purpose or end goals. Historically, it involved explaining the functions and adaptations of organisms by attributing purpose or design to them. While teleological explanations were common in early biology, the rise of modern science, particularly Darwinian evolution, challenged, and reshaped teleological thinking. Here's a detailed exploration of teleology in biology:

1. Historical Perspective:

- **Aristotelian Influence:** Teleological thinking traces back to the ancient Greeks, especially Aristotle, who believed that organisms were perfectly adapted to their specific functions, implying a purpose in their design.
- **Natural Theology:** In the 18th and early 19th centuries, natural theology argued that the complexity and adaptation seen in nature were evidence of a divine purpose or design, reflecting the work of a Creator.

2. Teleological Explanations:

- **Purposeful Adaptations:** Teleological explanations involve attributing specific adaptations of organisms to a purpose or goal, implying that these features exist because they serve a particular function.
- **Examples:** The human eye being adapted for vision, or the wings

of birds being designed for flight were explained teleologically in the past.

3. Challenges to Teleology:
    - **Darwinian Evolution:** Charles Darwin's theory of evolution by natural selection provided a naturalistic explanation for the diversity and adaptations of life. It replaced teleological explanations with the concept of gradual, non-purposeful evolution through natural processes.
    - **Random Variation and Selection:** Darwinian evolution relies on random variations in traits and natural selection, not guided by any purpose or goal. Adaptations arise through the differential survival and reproduction of individuals with traits suited to their environment.
4. Modern Perspectives:
    - **Functional Explanation:** In modern biology, explanations often focus on the function of traits without implying purpose. Organisms and their traits are studied in terms of how they contribute to survival, reproduction, and ecological interactions.
    - **Proximate vs. Ultimate Explanations:** Proximate explanations address how a trait functions at a physiological level, while ultimate explanations consider the evolutionary reasons for the trait's existence. Ultimate explanations are non-teleological, emphasizing historical processes rather than purpose.
5. Teleonomy and Systems Biology:
    - **Teleonomy:** It refers to apparent purpose or goal-directedness in biological systems. It acknowledges that organisms exhibit complex, coordinated behaviors and traits that may appear purposeful without invoking supernatural or divine explanations.
    - **Systems Biology:** The study of biological systems, including their emergent properties and network interactions, often explores the functional integration of components without necessarily implying purpose in a teleological sense.
6. Ethical and Philosophical Implications:
    - **Ethical Considerations:** Teleological thinking sometimes influences ethical debates, such as arguments about the purpose of certain biological traits or the perceived purpose in ecological relationships.

- **Philosophical Discussions:** Teleology remains a topic of philosophical discussion, exploring concepts of purpose, intentionality, and the apparent design in nature without invoking supernatural causes.

So, teleology in biology, once prevalent in explaining adaptations and functions, has been largely replaced by naturalistic, non-purposeful explanations rooted in evolutionary theory and modern scientific understanding. While apparent purposefulness in biological systems continues to fascinate scientists and philosophers, contemporary biological explanations emphasize natural processes, random variation, and selection without invoking supernatural or purpose-driven causes.

7.3. OVERVIEW OF PHILOSOPHY OF EVOLUTION

The philosophy of evolution encompasses a range of perspectives and inquiries related to the fundamental questions about life, existence, and the natural world. Here is an overview of key aspects within the philosophy of evolution:

1. Epistemological Questions:
    - **Scientific Methodology:** Philosophers explore the scientific methods used in evolutionary biology, considering issues like falsifiability, empirical evidence, and the formulation of hypotheses.
    - **Nature of Scientific Theories:** Evolutionary theory raises questions about the nature of scientific theories, such as the role of natural selection, genetic variation, and the integration of different biological disciplines.
2. Metaphysical Questions:
    - **Nature of Species:** Philosophers debate the concept of species in the context of evolution. Evolution challenges traditional definitions, leading to discussions about the continuity of species and the arbitrariness of species boundaries.
    - **Teleology and Purpose:** Evolutionary theory raises questions about teleology – whether evolution has a purpose or direction. Philosophers explore the implications of natural selection for the concept of purpose in the natural world.

3. Ontological Questions:

- **Origins of Life:** Philosophers examine questions about the origin of life and the transition from non-living to living matter. Evolutionary theory intersects with abiogenesis theories, exploring the boundary between the living and non-living.
- **Mind and Consciousness:** Evolutionary psychology raises questions about the evolution of the human mind and consciousness. Philosophers explore how evolutionary processes have shaped cognition, emotion, and self-awareness.

4. Ethical and Social Implications:

- **Ethical Challenges:** Evolutionary theory raises ethical questions, especially regarding human evolution, genetics, and animal welfare. Philosophers explore the ethical implications of genetic engineering, evolutionary psychology, and human-animal relationships.
- **Societal Impact:** Philosophers discuss the societal impact of evolutionary ideas, particularly in areas such as education, religion, and public policy. Debates often revolve around the compatibility of evolutionary theory with religious beliefs and its role in education curricula.

5. Historical and Philosophical Context:

- **Historical Development:** Philosophers analyze the historical development of evolutionary thought, including the contributions of Darwin, Wallace, and subsequent evolutionary biologists. They explore how societal, cultural, and religious contexts influenced the reception of evolutionary ideas.
- **Philosophical Critiques:** Evolutionary theory faces philosophical critiques, including questions about determinism, reductionism, and the limits of adaptation. Philosophers critically assess the assumptions and implications of evolutionary concepts.

6. Interdisciplinary Connections:

- **Philosophy of Biology:** Evolutionary theory is a central topic in the philosophy of biology. Philosophers explore biological concepts such as adaptation, natural selection, and species definitions within an evolutionary framework.
- **Philosophy of Science:** Evolutionary biology serves as a case study in the philosophy of science, raising questions about theory

confirmation, scientific explanation, and the demarcation problem between science and pseudoscience.

In essence, the philosophy of evolution delves into the foundational questions underlying evolutionary biology, exploring the nature of life, the mechanisms of evolution, and the broader implications for our understanding of the natural world and our place within it. Philosophical inquiry continues to enrich our conceptual framework, fostering deeper insights into the complex interplay between evolution, biology, and the broader philosophical inquiries about existence and knowledge.

REFERENCES

1. An anthology of the most philosophically important papers on evolution, including many otherwise inaccessible ones.
2. Axelrod, R., (1984). *The Evolution of Cooperation*. Basic Books.
3. Berry, R. J., (1988). *God and Evolution: Creation, Evolution, and the Bible*. Hodder and Stoughton.
4. Bonner, J. T., (1988). *The Evolution of Complexity by Means of Natural Selection*. Princeton University Press.
5. Bowler, P. J., (1983). *The Eclipse of Darwinism: Antievolutionary Theories in the Decades Around 1900*. Johns Hopkins.
6. Dawkins, R., & Krebs, J. R., (1979). Arms races within and between species. *Proc. R. Soc. Lond. B., 205*, 480–512.
7. Dawkins, R., (1976). *The Selfish Gene* (2nd edn.). Oxford University Press–1989.
8. Dawkins, R., (1995). *River Out of Eden*. Weidenfeld and Nicholson.
9. Dawkins, R., (1996). *Climbing Mount Improbable*. Viking Press.
10. Dennett, D., (1995). *Darwin's Dangerous Idea: Evolution and the Meanings of Life*. Allen Lane Press.
11. Depew, D. J., & Weber, B. H., (1995). *Darwinism Evolving: Systems Dynamics and the Genealogy of Natural Selection*. Bradford Books/ MIT Press.
12. Dray, W. H., (1957). *Laws and Explanation in History*. Oxford University Press.
13. Dray, W. H., (1966). *Philosophical Analysis and History*. Harper and Row.
14. Duhem, P., (1914). *The Aim and Structure of Physical Theory*. Princeton University Press (English trans. 1954).
15. Ereshefsky, M., (1991). The semantic approach to evolutionary theory. *Biology and Philosophy*, *6*, 59–80.
16. Feyerabend, P. K., (1970a). Consolations for the specialist. In: Lakatos, I., & Musgrave, I., (eds.), *Criticism*, and *the Growth of Knowledge*. Cambridge University Press.
17. Feyerabend, P. K., (1970b). Against method. In: *Minnesota Studies in Philosophy* (Vol. 4).
18. Feyerabend, P. K., (1975). *Against Method*. Verso Editions.

19. Flew, A., & MacIntyre, A., (1955). *New Essays in Philosophical Theology*. SCM Press.
20. Franklin, J., (1997). Stove's anti-Darwinism. *Philosophy*, *72*, 133–136.
21. Gayon, J., (1996). The individuality of the species: A Darwinian theory? from Buffon to Ghiselin, and back to Darwin. *Biology and Philosophy*, *11*, 215–244.
22. Ghiselin, M. T., (1975). A radical solution to the species problem. *Systematic Zoology*, *23*, 536–544.
23. Gould, S. J., & Lewontin, R., (1979). The spandrels of San Marco and the Panglossian paradigm: A critique of the adaptationist program. *Proc. R. Soc. Lond. B.*, *205*, 581–598.
24. Gould, S. J., (1989). *Wonderful Life: The Burgess Shale and the Nature of History*. Norton.
25. Gould, S. J., (1992). *Kropotkin was No Crackpot* (p. 325). Essay 13 in Bully for Brontosaurus, Penguin.
26. Gould, S. J., (1996). *Full House*. Harmony, published outside the US as Life's Grandeur: The Spread of Excellence from Plato to Darwin, Random House.
27. Griffiths, P. E., (1996). The historical turn in the study of adaptation. *British Journal for the Philosophy of Science*, *47*, 511–532.
28. Hull, D., (1974). *The Philosophy of Biological Science*. Prentice-Hall.
29. Hull, D., (1976). Are species really individuals? *Systematic Zoology*, *25*, 174–191.
30. Hull, D., (1988). *Science as a Process: An Evolutionary Account of the Social and Conceptual Development of Science*. University of Chicago Press.
31. Hull, D., (1989). *The Metaphysics of Evolution*. State University of New York Press.
32. Kripke, S., (1972). Naming and necessity. In: Davidson, D., & Harman, G., (eds.), *Semantics and Natural Language* (pp. 253–355), Reidel.
33. Kuhn, T. S., (1962). *The Structure of Scientific Revolutions* (2nd edn., 1970). University of Chicago Press.
34. Kuhn, T. S., (1970). Reflections on my critics. In: Lakatos, I., & Musgrave, I., (eds.), *Criticism and the Growth of Knowledge*. Cambridge University Press.
35. Kuhn, T. S., (1972). Second thoughts on paradigms. In: Suppe, F., (ed.),

The Structure of Scientific Theories (2nd edn.). University of Illinois Press–1977.

36. Lakatos, I., (1970). Falsification and the methodology of scientific research programs. In: Lakatos, I., & Musgrave, I., (eds.), *Criticism and the Growth of Knowledge*. Cambridge University Press.
37. Laudan, L., (1977). *Progress and Its Problems*. University of California Press.
38. Lennox, J. G., (1992). Teleology. In: Keller, E. F., & Lloyd, E. A., (eds.), *Keywords in Evolutionary Biology*. Harvard University Press.
39. Lurie, E., (1988). *Louis Agassiz: A Life in Science*. Johns Hopkins University Press.
40. Masterman, M., (1970). The nature of paradigms. In: Lakatos, I., & Musgrave, I., (eds.), *Criticism and the Growth of Knowledge*. Cambridge University Press.
41. Mayr, E., (1970). *Populations, Species, and Evolution*. Harvard University Press.
42. Mayr, E., (1982). *The Growth of Biological Thought: Diversity, Evolution, and Inheritance*. Belknap Press/Harvard University Press.
43. Mayr, E., (1988). *Toward a New Philosophy of Biology: Observations of an Evolutionist*. Belknap Press/Harvard University Press.
44. Miller, K., (1982). Answers to standard creationist arguments. *Creation/Evolution*, *3*, 1–13.
45. Monod, J., (1972). *Chance and Necessity*. Collins.
46. Nagel, E., (1961). *The Structure of Science: Problems in the Logic of Scientific Explanation*. Routledge and Kegan Paul.
47. O'Grady, D., & Brooks, D., (1988). Teleology and biology. In: Weber, B. H., Depew, D. J., & Smith, J. D., (eds.), *Entropy, Information, and Evolution: New Perspectives on Physical and Biological Evolution*. MIT Press.
48. Oldroyd, D., (1986). *The Arch of Knowledge: An Introductory Study of the History of the Philosophy and Methodology of Science*. Methuen.
49. Panchen, A. L., (1992). *Classification, Evolution, and the Nature of Biology*. Cambridge University Press.
50. Popper, K., (1976). *Unended Quest: An Intellectual Autobiography*. Fontana Press.
51. Putnam, H., (1975). *Mind, Language, and Reality*. Cambridge University Press.

52. Quine, W. V. O., (1969). Natural kinds. In: *Ontological Relativity and Other Essays*. Columbia University Press.
53. Ruse, M., (1979). *The Darwinian Revolution: Science Red in Tooth and Claw*. University of Chicago Press.
54. Ruse, M., (1989). *The Darwinian Paradigm: Essays on its History, Philosophy, and Religious Implications*. Routledge.
55. Ruse, M., (1992). Darwinism. In: Keller, E. F., & Lloyd, E. A., (eds.), *Keywords in Evolutionary Biology*. Harvard University Press.
56. Ruse, M., (1997). *Monad to Man: The Concept of Progress in Evolutionary Biology*. Harvard University Press.
57. Selkirk, D. R., & Burrows, F. J., (1987). *Confronting Creationism: Defending Darwin*. New South Wales University Press.
58. Sober, E., (1984). *The Nature of Selection: Evolutionary Theory in Philosophical Focus*. Bradford Books/MIT Press.
59. Sober, E., (1988). *Reconstructing the Past: Parsimony, Evolution, and Inference*. Bradford Books/MIT Press.
60. Sober, E., (1994). *Conceptual Issues in Evolutionary Biology* (2nd edn.). Cambridge University Press.
61. Stamos, J., (1996). Popper, falsifiability, and evolutionary biology. *Biology and Philosophy*, *11*, 161–191.
62. Sterelny, K., (1995). Understanding life: Recent work in the philosophy of biology. *Brit. J. Phil. Sci., 46*, 155–183.
63. Stove, D., (1995). *Darwinian Fairytales*. Avebury Press.
64. Suppe, F., (1977). *The Structure of Scientific Theories* (2nd edn.). University of Illinois Press.
65. Suppe, F., (1989). *The Semantic Conception of Theories and Scientific Realism*. University of Illinois Press.
66. The standard reference on the philosophical implications of natural selection.
67. The title says it all, really. A first-class introduction to the state of play.
68. This is a readable and comprehensive introduction to the issues of epistemology over the last 2500 years.
69. Vermeij, G. J., (1987). *Evolution and Escalation: An Ecological History of Life*. Princeton University Press.
70. Weber, B. H., & Depew, D. J., (1996). *Natural Selection and Self-Organization, Biology, and Philosophy*, *11*, 33–65.

71. Williams, G. C., (1966). *Adaptation and Natural Selection*. Princeton University Press, reissued with a new Introduction, 1996.
72. Williams, G. C., (1985). In: Dawkins, R., & Ridley, M., (eds.), *Defense of Reductionism*. Oxford Studies in Evolutionary Biology, Oxford University Press.
73. Williams, G. C., (1992). *Natural Selection: Domains, Levels, and Challenges*. Oxford University Press.

CHAPTER 8

Mechanisms of Evolution

CONTENTS

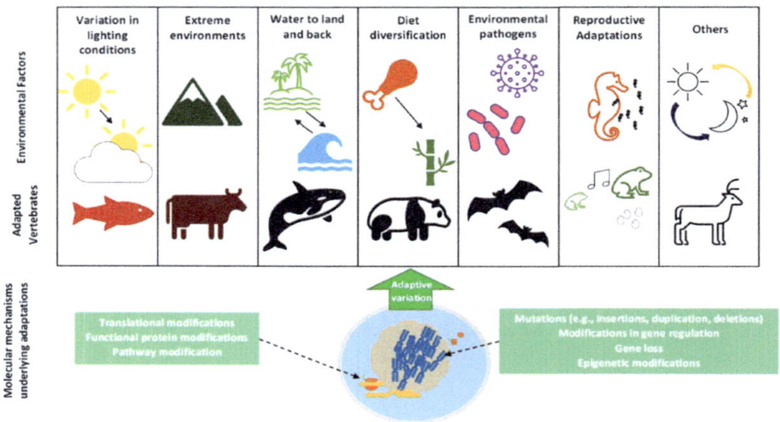

Figure 8.1. *Mechanisms of evolution.*

Source: Martínez, S. F., & Pilot, M., (2023). Molecular mechanisms underlying vertebrate adaptive evolution: A systematic review. Genes, 14(2), 416. https://doi.org/10.3390/genes14020416; Creative Commons License.

8.1. NATURAL SELECTION

Natural selection is a fundamental mechanism of evolution, first described by Charles Darwin, which explains how species adapt to their environment over time. It is a process by which certain heritable traits become more or less common in a population based on their impact on an organism's ability to survive and reproduce. Here's a detailed explanation of natural selection:

1. Variation Within Populations:
   - i. **Inherent Diversity:** Within any population, individuals exhibit variation in their traits, such as size, color, shape, or behavior. This variation is often heritable, meaning it is passed from parents to offspring through genes.
2. Struggle for Existence:
   - i. **Limited Resources:** Resources in the environment, such as food, water, and shelter, are limited. Consequently, individuals within a population must compete for these resources to survive and reproduce.
   - ii. **Predation and Disease:** Additionally, populations face challenges from predators, diseases, and other environmental factors that can reduce an individual's chances of survival.

3. Differential Survival and Reproduction:
   i. **Natural Selection at Work:** Not all individuals in a population survive and reproduce to their full potential due to these challenges. Certain traits provide advantages in the struggle for existence. Individuals with advantageous traits are more likely to survive, reproduce, and pass on these traits to their offspring.
   ii. **Selective Pressure:** The environment acts as a selective pressure, favoring traits that enhance an organism's survival and reproduction. These traits increase the individual's fitness, defined as its ability to survive, and reproduce in its environment.
4. Heritability of Traits:
   i. **Transmission of Traits:** Crucially, the advantageous traits that lead to higher fitness must be heritable. If a particular trait that provides an advantage is heritable, it will be passed on to the next generation.
   ii. **Genetic Basis:** Traits are often determined by genes, and if these traits increase an individual's chances of survival and reproduction, the genes responsible for these traits become more common in the population over generations.
5. Accumulation of Adaptations:
   i. **Gradual Changes:** Over long periods of time, natural selection leads to the accumulation of adaptations in populations. Adaptations are traits or characteristics that increase an organism's fitness in its environment. As advantageous traits become more common in a population, the population becomes better suited to its environment, reflecting the gradual process of adaptation through natural selection.
6. Role of Environmental Change:
   i. **Dynamic Process:** Natural selection is dynamic and responsive to environmental changes. When the environment changes, certain traits that were advantageous might become less so, leading to the selection of new traits that are better suited to the altered environment. This process allows populations to adapt to changing conditions over time.

7. Examples of Natural Selection:
   i. **Industrial Melanism:** One classic example of natural selection is the case of industrial melanism in peppered moths during the Industrial Revolution. Dark-colored moths became more prevalent in polluted environments, as they were better camouflaged against soot-covered trees, illustrating the influence of natural selection on camouflage traits.
   ii. **Antibiotic Resistance:** Another well-known example is the evolution of antibiotic resistance in bacteria. Due to the overuse of antibiotics, bacteria with resistance genes have a higher chance of survival and reproduction. Consequently, antibiotic-resistant strains become more common in bacterial populations.

So, natural selection is a fundamental and powerful process that drives the evolution of species over time. It explains how organisms become adapted to their environments and how new species can arise, leading to the remarkable diversity of life observed on Earth. Natural selection, along with other mechanisms of evolution, forms the foundation of modern evolutionary biology, providing a deep understanding of the complexities of life's development and diversification.

8.2. GENETIC VARIATION

Genetic variation refers to the differences in genetic material within a population or among individuals of the same species. It is a fundamental concept in biology and a key driver of evolution. Here's a detailed explanation of genetic variation:

1. Origin of Genetic Variation:
   i. **Mutation:** These are random changes in an organism's DNA sequence. They can occur due to errors during DNA replication or under the influence of external factors like radiation or chemicals. Mutations are the ultimate source of genetic variation and can create new alleles (alternative forms of genes) in a population.
   ii. **Genetic Recombination:** During sexual reproduction, genetic material from two parents is combined, creating offspring with a unique combination of genes. This

process, known as genetic recombination, occurs during the formation of gametes (sperm and egg cells) and introduces further genetic diversity.

2. Types of Genetic Variation:
   i. **Single Nucleotide Polymorphisms (SNPs):** These are the most common type of genetic variation. They involve a single nucleotide change in the DNA sequence. SNPs can occur within coding or non-coding regions of genes and may influence traits or disease susceptibility.
   ii. **Insertions and Deletions (Indels):** Indels involve the insertion or deletion of nucleotides in a DNA sequence. Depending on their location within genes, they can lead to frameshift mutations, altering the amino acid sequence of proteins.
   iii. **Copy Number Variations (CNVs):** These are large segments of DNA that are duplicated or deleted. CNVs can affect gene dosage, potentially leading to phenotypic differences.
   iv. **Structural Variations:** These include larger-scale alterations, such as inversions, translocations, or chromosome number variations. Structural variations can have significant effects on gene expression and function.
3. Importance of Genetic Variation:
   i. **Basis of Evolution:** Genetic variation is the raw material upon which natural selection acts. Individuals with advantageous genetic variations are more likely to survive and reproduce, passing these variations to the next generation. Over time, this leads to the adaptation and evolution of populations.
   ii. **Disease Resistance:** Genetic variation in populations provides the basis for resistance to diseases. Some individuals may have genetic variants that confer resistance to specific pathogens, allowing them to survive and reproduce in disease-prone environments.
   iii. **Environmental Adaptations:** Genetic variation allows populations to adapt to different environments. Traits that provide advantages in specific environments, such as heat

tolerance or resistance to harsh climates, can become more common due to natural selection.

4. Measuring Genetic Variation:
   i. **Population Genetics:** It is the study of genetic variation within populations. Scientists use tools like allele frequency analysis and genetic markers to measure genetic diversity and study how genes are distributed within and between populations.
   ii. **Genetic Mapping:** These techniques, including linkage mapping and association studies, help identify specific genetic variants associated with traits, diseases, or other phenotypic characteristics.
5. Maintaining Genetic Variation:
   i. **Balancing Selection:** Certain selective pressures, such as heterozygote advantage or frequency-dependent selection, can maintain genetic variation in populations. Heterozygote advantage occurs when individuals with two different alleles have higher fitness, ensuring the persistence of both alleles in the population.
   ii. **Gene Flow:** The movement of genes between populations, can introduce new genetic variation and prevent populations from becoming genetically isolated. It maintains genetic diversity by reducing genetic differentiation between populations.

Therefore, genetic variation is a central concept in genetics and evolutionary biology. It provides the raw material for natural selection, allowing populations to adapt to changing environments and ensuring the diversity of life on Earth. Understanding genetic variation is crucial for fields such as medicine, agriculture, and conservation biology, as it influences disease susceptibility, crop breeding, and the preservation of endangered species.

8.3. GENETIC DRIFT

Genetic drift is a mechanism of evolution that describes the random fluctuations of allele frequencies within a population from generation to generation. Unlike natural selection, genetic drift does not involve adaptation to the environment; instead, it arises from chance events. Here's a detailed explanation of genetic drift:

1. Random Sampling and Allele Frequencies:
   - i. **Small Population Sizes:** Genetic drift is particularly influential in small populations where chance events can have a significant impact. In small populations, the genetic composition can change dramatically over generations due to random sampling during reproduction.
   - ii. **Allele Frequencies:** The proportion of a particular allele in a population, can fluctuate purely by chance. When a small number of individuals reproduce, the alleles they carry may not be representative of the overall population's genetic makeup. This discrepancy can lead to shifts in allele frequencies in the next generation.
2. Founder Effect and Population Bottlenecks:
   - i. **Founder Effect:** When a small group of individuals establishes a new population, the genetic diversity of the new population may be drastically reduced compared to the original population. The genetic makeup of the founder group can heavily influence the genetic composition of the entire population, leading to unique allele frequencies.
   - ii. **Population Bottleneck:** It occurs when a large population is drastically reduced in size, often due to a natural disaster or human intervention. The surviving individuals may not represent the genetic diversity of the original population, leading to a bottleneck effect where only a subset of alleles is passed to the next generation.
3. Fixation and Loss of Alleles:
   - i. **Fixation:** Genetic drift can lead to the fixation of an allele, where it becomes the only allele present in the population. This occurs when, by chance, all other alleles are lost, especially in small populations.
   - ii. **Loss of Alleles:** Conversely, genetic drift can also lead to the loss of alleles. If a rare allele is not passed on to the next generation due to random chance, it may be lost from the population entirely.

4. Effects on Genetic Diversity:
   i. **Reduced Genetic Diversity:** Genetic drift tends to reduce genetic diversity within populations over time. In small populations, specific alleles can become fixed, and the overall diversity of the population decreases.
   ii. **Increased Genetic Differences between Populations:** Genetic drift can also increase genetic differences between populations. If different populations experience different random events, they can diverge genetically even without environmental differences driving natural selection.
5. Genetic Drift vs. Natural Selection:
   i. **Random vs. Adaptive:** Genetic drift is entirely random, while natural selection is adaptive, meaning it acts on traits that affect an organism's survival and reproduction. Genetic drift does not lead to adaptations; it merely changes allele frequencies due to chance events.
6. Long-Term Effects and Evolutionary Significance:
   i. **Neutral Theory:** In population genetics, the neutral theory suggests that many genetic changes are selectively neutral, meaning they neither confer advantages nor disadvantages in terms of fitness. Neutral changes can be driven by genetic drift rather than natural selection.
   ii. **Evolutionary Consequences:** While genetic drift alone may not lead to adaptations, it can have significant long-term effects on populations. In conjunction with other evolutionary forces like natural selection and gene flow, genetic drift contributes to the evolution of species and their adaptations to specific environments.

In summary, genetic drift is a crucial mechanism in the evolution of populations. It emphasizes the role of chance events in shaping allele frequencies, highlighting the stochastic nature of evolution. While natural selection is a powerful force driving adaptation, genetic drift plays a significant role in small populations and can lead to genetic divergence and the fixation or loss of alleles within populations over time.

8.4. GENE FLOW

Gene flow, also known as migration or allele flow, is the transfer of genetic material from one population to another within the same species. This movement of genes occurs when individuals or gametes (sperm and egg cells) migrate and breed with individuals from another population. Gene flow plays a crucial role in shaping the genetic diversity and evolutionary trajectories of populations. Here's a detailed explanation of gene flow:

1. Migration of Individuals:
   - i. **Physical Movement:** Gene flow occurs when individuals migrate from one population to another and successfully reproduce with members of the new population.
   - ii. **Geographical Barriers:** These like mountains, rivers, or oceans can restrict gene flow between populations. When individuals or their gametes cross these barriers, they introduce new genetic material to a previously isolated population.
2. Gene Flow and Genetic Diversity:
   - i. **Increase in Genetic Diversity:** Gene flow increases genetic diversity within populations. New genetic variants introduced by migrants can increase the variety of alleles within a population, enhancing its ability to adapt to changing environments.
   - ii. **Homogenization between Populations:** Gene flow can homogenize the genetic makeup of different populations. When individuals from different populations interbreed, genetic differences between populations decrease, making them more similar genetically.
3. Preventing Genetic Differentiation:
   - i. **Counteracting Genetic Differentiation:** Gene flow counteracts the genetic differentiation that arises due to other evolutionary forces like genetic drift or natural selection. It helps maintain genetic cohesion within species.
   - ii. **Preservation of Species Identity:** Gene flow is crucial in preventing isolated populations from becoming reproductively isolated, which can eventually lead to speciation (the formation of new species).
4. Factors Influencing Gene Flow:
   - i. **Distance and Mobility:** Gene flow is often inversely

proportional to the distance between populations. More mobile species have higher gene flow rates.

ii. **Barriers:** Physical barriers, such as mountains or deserts, can impede gene flow. However, some species have adaptations or behaviors that allow them to overcome these barriers, facilitating gene flow.

iii. **Mate Choice:** This and reproductive preferences can influence gene flow. Individuals might prefer mating with individuals from their own population, limiting gene flow with other populations.

5. Genetic Consequences:

   i. **Introduction of Beneficial Traits:** Gene flow can introduce beneficial traits, such as resistance to diseases or environmental adaptations, into populations. These traits can spread rapidly through the population via gene flow.

   ii. **Dilution of Adaptations:** Gene flow from populations with different selective pressures can dilute locally adapted traits, potentially reducing the fitness of certain populations in their specific environments.

6. Human-Mediated Gene Flow:

   i. **Crop Cultivation and Animal Breeding:** Humans have facilitated gene flow in plants and animals through agricultural practices and selective breeding. This has led to the spread of desirable traits in domesticated species.

   ii. **Impact on Wild Populations:** Human activities can also lead to unintended gene flow between wild and domesticated populations, influencing the genetic diversity of natural ecosystems.

In fact, gene flow is a fundamental process in evolution that maintains genetic diversity within and between populations. It prevents genetic divergence, aids in adaptation to changing environments, and plays a significant role in the preservation of species integrity. Understanding the patterns and mechanisms of gene flow is essential for studying the evolution of populations and species.

8.5. MUTATION

Mutation is a fundamental process in biology that introduces genetic variation, serving as the raw material for evolution. It involves changes in the DNA sequence of an organism, leading to alterations in the inherited traits passed from one generation to the next. Here's a detailed explanation of mutations:

1. Types of Mutations:

i. Point Mutations: These involve changes in a single nucleotide of the DNA sequence. They can be categorized into three types:

- **Substitution:** One nucleotide is replaced by another.
- **Insertion:** One or more nucleotides are added to the sequence.
- **Deletion:** One or more nucleotides are removed from the sequence.

ii. Chromosomal Mutations: These mutations involve changes in the structure or number of whole chromosomes. Examples include:

- **Duplication:** A section of the chromosome is duplicated.
- **Deletion:** A section of the chromosome is lost.
- **Inversion:** A section of the chromosome is reversed.

d. Translocation: A segment of one chromosome is transferred to another chromosome.

2. Causes of Mutations:

i. **Spontaneous Mutations:** These occur naturally due to errors during DNA replication or repair. These errors can be caused by environmental factors or intrinsic cellular processes.

ii. **Induced Mutations:** These result from exposure to external factors such as radiation, chemicals, or certain biological agents. These mutagens can increase the mutation rate.

3. Effects of Mutations:

i. **Silent Mutations:** Some mutations do not result in changes to the amino acid sequence of a protein. These silent mutations occur in non-coding regions of genes or code for the same amino acid.

ii. **Missense Mutations:** These change one amino acid in the protein sequence, potentially altering its structure and function.

iii. **Nonsense Mutations:** These create a premature stop codon in the mRNA, leading to a truncated, nonfunctional protein.

iv. **Frameshift Mutations:** These result from insertions or deletions of nucleotides, altering the reading frame of the gene and producing a completely different amino acid sequence.

v. **Gain-of-Function and Loss-of-Function Mutations:** Some mutations enhance the function of a protein (gain-of-function), while others reduce or abolish its function (loss-of-function).

4. Role in Evolution:

i. **Genetic Diversity:** Mutations create genetic diversity within populations. New alleles generated by mutations can lead to different traits, potentially affecting an organism's survival and reproductive success.

ii. **Natural Selection:** Mutations provide the variation upon which natural selection acts. Beneficial mutations can be favored, leading to their increase in the population, while deleterious mutations can be eliminated.

iii. **Neutral Theory:** According to the neutral theory of molecular evolution, many mutations are selectively neutral, meaning they do not significantly impact an organism's fitness. Neutral mutations can accumulate over time due to genetic drift.

5. Applications and Importance:

i. **Genetic Diseases:** Mutations can cause genetic diseases in humans and other organisms. Understanding the genetic basis of these diseases is crucial for medical research and treatment.

ii. **Biotechnology:** Scientists can induce specific mutations in organisms to study gene function, develop genetically modified organisms, or produce desired traits in crops or livestock.

iii. **Evolutionary Studies:** Mutations and their effects provide valuable insights into the evolutionary history of species and the processes that drive speciation and adaptation.

Thus, mutations are essential biological phenomena that drive genetic diversity, evolution, and the adaptation of organisms to changing environments. They play a central role in the process of evolution, enabling species to evolve and adapt over time. Studying mutations is critical for understanding both the molecular basis of genetic diseases and the broader mechanisms that underpin the diversity of life on Earth.

8.6. SEXUAL SELECTION

Sexual selection is a specific type of natural selection that occurs when individuals within a population differ in their ability to attract mates, leading to the evolution of traits that enhance mating success. These traits might not necessarily provide a survival advantage but can greatly influence an individual's reproductive fitness. Here's a detailed explanation of sexual selection:

1. Types of Sexual Selection:
   i. **Intrasexual Selection:** Also known as competition between individuals of the same sex, occurs when members of one sex (usually males) compete with each other for access to individuals of the opposite sex (usually females). This competition can involve direct physical combat, displays, or other forms of rivalry.
   ii. **Intersexual Selection:** Also known as mate choice, occurs when members of one sex (usually females) choose their mates based on specific traits. These traits might include physical characteristics, behaviors, or other attributes that indicate good genes, health, or parental qualities.
2. Male-Male Competition:
   i. **Combat:** In many species, males engage in physical combat to establish dominance and access to females. Antlers in deer or horns in beetles are examples of traits evolved through male-male competition.
   ii. **Territorial Displays:** Males might defend territories rich in resources (such as nesting sites or food) that attract females. Displays, songs, or other behaviors are used to deter rival males.
3. Female Choice:
   i. **Ornamental Traits:** Males often possess elaborate and colorful traits, such as bright plumage in birds or vibrant fins in fish. These traits are favored by females and indicate the male's genetic quality.
   ii. **Courtship Displays:** Males may perform intricate courtship displays, showcasing their abilities, stamina, or genetic fitness. Females select mates based on the quality of these displays.
4. Parental Investment Theory:
   i. **Anisogamy:** Sexual selection is closely related to anisogamy, the phenomenon where males and females produce different-sized gametes (smaller sperm and larger eggs). As females invest more

in reproduction (due to gestation, nursing, etc.), they are generally more selective in mate choice.

ii. **Bateman's Principle:** It states that in species where females invest more in offspring, males typically compete for mating opportunities, while females are choosier about their mates.

5. Sexual Dimorphism:

i. **Sexual Ornaments:** Sexual selection often leads to sexual dimorphism, where males and females of the same species exhibit different physical traits. Males might have brighter colors or larger ornaments, while females have more subdued traits.

ii. **Trade-Offs:** Sexual dimorphism can result from trade-offs between traits. For instance, larger and more ornate ornaments might attract more mates but can also make males more vulnerable to predators.

6. Evolutionary Outcomes:

i. **Runaway Selection:** Runaway sexual selection occurs when a preference for a specific trait in one sex leads to an exaggerated development of that trait. This positive feedback loop can result in extreme and diverse physical characteristics within a population.

ii. **Sensory Bias:** Sometimes, traits evolve not because they directly benefit the individual, but because they exploit pre-existing sensory biases in the opposite sex. These biases can lead to the evolution of specific traits that appeal to the sensory preferences of potential mates.

So, sexual selection is a powerful evolutionary force shaping the physical and behavioral traits of organisms. It often results in elaborate and sometimes costly traits that enhance reproductive success. By driving the evolution of these traits, sexual selection plays a crucial role in the diversification and adaptation of species over generations.

8.7. SPECIATION

Speciation is the process by which new biological species arise. It is a fundamental mechanism in evolutionary biology, marking the point at which a single evolutionary lineage splits into two or more distinct species that can no longer interbreed and produce fertile offspring. Here's a detailed explanation of speciation:

1. Isolation Mechanisms:
   i. **Geographical Isolation:** Geographical barriers such as mountains, rivers, or oceans can physically separate populations, preventing them from interbreeding. Over time, isolated populations can evolve different traits due to independent evolutionary pressures.
   ii. **Reproductive Isolation:** These mechanisms prevent members of different populations from mating successfully. There are two main types:
   iii. **Prezygotic Isolation:** These mechanisms prevent the formation of viable hybrids before fertilization. Examples include differences in mating rituals, incompatible genitalia, or differences in breeding seasons.
   iv. **Postzygotic Isolation:** Postzygotic mechanisms prevent the development or reproduction of hybrid offspring. Hybrid inviability, where hybrids do not survive to adulthood, and hybrid sterility, where hybrids are infertile, are examples of postzygotic isolation.
2. Modes of Speciation:
   i. **Allopatric Speciation:** This occurs when populations become geographically isolated. Over time, different environments and selective pressures lead to genetic divergence between the isolated populations, eventually making them distinct species.
   ii. **Sympatric Speciation:** This occurs within the same geographic area. It often involves the development of different ecological niches or specific mating preferences within the same population, leading to reproductive isolation.
   iii. **Parapatric Speciation:** This occurs when populations are geographically adjacent but have limited gene flow due to specific localized adaptations or preferences. Gradual changes along environmental gradients can lead to the evolution of distinct species.
3. Genetic Divergence:
   i. **Accumulation of Genetic Differences:** Isolated populations accumulate genetic differences over time due

to mutations, genetic drift, and natural selection acting independently on each population.

ii. **Divergent Selection:** Different selective pressures in different environments can lead to divergent selection, where traits that are advantageous in one environment are disadvantageous in another. This divergence contributes to the evolution of distinct species.

4. Hybridization and Speciation:

i. **Hybrid Zones:** In some cases, previously isolated species come into contact, leading to hybridization. Hybrid zones are regions where genetically distinct species interbreed, potentially leading to the formation of new hybrid species or reinforcing reproductive barriers between the parent species.

5. Evolutionary Consequences:

i. **Biodiversity:** Speciation is a major driver of biodiversity, leading to the formation of numerous species with diverse traits and adaptations.

ii. **Adaptation to Environment:** Speciation allows populations to adapt to specific environments, leading to the evolution of species suited to different ecological niches.

iii. **Ecological Relationships:** Speciation can lead to the development of complex ecological relationships, including predator-prey dynamics, mutualistic interactions, and coevolution between species.

6. Human Influence on Speciation:

i. **Anthropogenic Speciation:** Human activities, such as habitat fragmentation, introduction of species to new environments, and selective breeding, can lead to new species formation, often at an accelerated rate compared to natural processes.

In summary, speciation is a complex and dynamic process driven by genetic and environmental factors. It results in the formation of new species, contributing to the rich diversity of life on Earth. Understanding the mechanisms of speciation is crucial for evolutionary biology, ecology, and conservation efforts, helping scientists comprehend the origins and relationships of various species in the natural world.

8.8. ADAPTIVE RADIATION

Adaptive radiation is a process in evolutionary biology where a single ancestral species rapidly diversifies into a variety of forms to adapt to different environments or ecological niches. This phenomenon results in the evolution of multiple species, each uniquely adapted to specific habitats and lifestyles. Here's a detailed explanation of adaptive radiation:

1. Key Features of Adaptive Radiation:
    i. **Rapid Diversification:** Adaptive radiation involves the rapid divergence of a lineage into multiple species, often in a relatively short geological time span.
    ii. **Ecological Opportunity:** Adaptive radiation typically occurs when organisms encounter new and diverse ecological opportunities, such as vacant niches or new habitats devoid of competitors.
    iii. **Divergent Evolution:** Species in adaptive radiations evolve distinct traits and adaptations to exploit different resources, leading to significant morphological, behavioral, and ecological diversity among closely related species.
2. Mechanisms and Factors Driving Adaptive Radiation:
    i. **Environmental Changes:** Significant environmental changes, such as the availability of new habitats, climate shifts, or the colonization of new geographic regions, can trigger adaptive radiation events.
    ii. **Ecological Niches:** When various ecological niches are unoccupied or available resources are underutilized, organisms can evolve specialized adaptations to exploit these niches, leading to the formation of new species.
    iii. **Key Innovations:** The evolution of novel traits or adaptations, known as key innovations, can enable organisms to access previously unavailable resources, promoting diversification within a lineage.
    iv. **Hybridization and Polyploidy:** Hybridization between different species or polyploidy (duplication of entire sets of chromosomes) can create genetic diversity, providing raw material for adaptive evolution.
3. Examples of Adaptive Radiation:

i. **Darwin's Finches:** The finches of the Galápagos Islands represent one of the most iconic examples of adaptive radiation. Different species of finches evolved unique beak shapes and sizes based on the type of food available on their respective islands.

ii. **Hawaiian Honeycreepers:** The Hawaiian Islands have been a hotspot for adaptive radiation, with honeycreepers evolving diverse bill shapes and sizes to exploit various food sources, including nectar, seeds, and insects.

iii. **Cichlid Fishes:** This in African rift lakes have undergone adaptive radiation, resulting in a wide array of species with specialized feeding habits and morphological traits adapted to different ecological niches.

4. Evolutionary Significance:

i. **Biodiversity:** Adaptive radiation contributes significantly to biodiversity, leading to the formation of numerous species within a relatively short period.

ii. **Scientific Understanding:** Studying adaptive radiation provides valuable insights into evolutionary processes, ecological interactions, and the factors driving speciation.

iii. **Conservation Implications:** Understanding adaptive radiation helps in devising conservation strategies, especially in areas with high biodiversity, by preserving key habitats and protecting species with unique adaptations.

In fact, adaptive radiation exemplifies the remarkable ability of organisms to diversify rapidly in response to new environmental opportunities. This process has played a crucial role in shaping the vast diversity of life on Earth and continues to be a subject of extensive research in the field of evolutionary biology.

8.9. CO-EVOLUTION

Co-evolution is a biological phenomenon where two or more species reciprocally influence each other's evolution. This interaction often leads to specific adaptations in each species, driven by the selective pressures imposed by the other species. Co-evolutionary relationships can occur between various pairs of organisms, such as predator-prey, host-parasite, and mutualistic interactions. Here's a detailed explanation of co-evolution:

1. **Types of Co-Evolution:**
   i. **Predator-Prey Co-Evolution:** Predators and their prey engage in co-evolutionary arms races. Predators evolve new hunting strategies or physical traits to catch prey more effectively, while prey species develop defenses and escape mechanisms to avoid predation.
   ii. **Host-Parasite Co-Evolution:** Host organisms and their parasites engage in co-evolution as hosts develop immune responses to combat parasites, and parasites evolve mechanisms to evade host defenses.
   iii. **Mutualistic Co-Evolution:** Mutualistic relationships, where two species benefit from each other (such as pollinators and flowering plants), can lead to co-evolution. Pollinators evolve traits to access resources, and plants develop traits to attract specific pollinators.
2. **Examples of Co-Evolution:**
   i. **Coevolution in Plants and Pollinators:** Flowering plants and their pollinators, such as bees or butterflies, have co-evolved specific traits. Flowers develop colors, shapes, and scents that attract certain pollinators, while pollinators, in turn, have evolved specialized mouthparts and behaviors to access nectar and pollen.
   ii. **Antagonistic Co-Evolution between Predators and Prey:** Prey species, such as gazelles and cheetahs, continually evolve adaptations for escape, while predators develop enhanced hunting abilities, resulting in a constant evolutionary arms race.
   iii. **Host-Parasite Co-Evolution in Bacteria and Immune Systems:** Bacteria and hosts engage in co-evolutionary battles. Bacteria evolve resistance to antibiotics, and hosts evolve new immune responses, leading to ongoing adaptations and counter-adaptations.
3. **Escalation and Red Queen Hypothesis:**
   i. **Escalation:** Co-evolution can escalate as each species develops counterstrategies against the other. For example, prey evolve better defenses, and predators develop more efficient hunting techniques, resulting in a continuous cycle

of adaptation and counter-adaptation.

ii. **Red Queen Hypothesis:** This proposes that species must continually evolve just to maintain their relative fitness in the face of ever-evolving species with which they interact. This term comes from Lewis Carroll's "Through the Looking-Glass," where the Red Queen tells Alice that she must run as fast as she can just to stay in the same place.

4. **Importance and Implications:**

i. **Biodiversity:** Co-evolution drives biodiversity by promoting the development of diverse traits and adaptations in interacting species.

ii. **Ecosystem Stability:** Co-evolutionary relationships can enhance the stability and functioning of ecosystems by promoting specialized interactions.

iii. **Human Impact:** Human activities, such as the use of antibiotics and pesticides, can disrupt co-evolutionary relationships, leading to ecological imbalances and potential consequences for biodiversity and agriculture.

In summary, co-evolution highlights the intricate and dynamic relationships between species, showcasing the constant adaptations and counter-adaptations that shape the diversity and complexity of life on Earth. Studying co-evolutionary interactions provides valuable insights into the evolutionary processes driving ecological relationships and biodiversity.

8.10. OVERVIEW OF MECHANISMS OF EVOLUTION

Evolution, the process by which species change over time, is driven by several key mechanisms. These mechanisms underpin the incredible diversity of life on Earth and are central to the study of biology. Here's a comprehensive analysis of the mechanisms of evolution:

1. Natural Selection:

- **Definition:** Natural selection is the process by which heritable traits that enhance an organism's survival and reproduction become more common in a population over successive generations.

Key Points:

- Individuals within a population exhibit variation in traits.
- Some traits provide advantages (adaptations) in the struggle for

survival and reproduction.

- Individuals with advantageous traits are more likely to survive, reproduce, and pass on these traits to the next generation.
- Natural selection results in the adaptation of populations to their environments over time.

2. Genetic Variation:

- **Definition:** Genetic variation refers to the diversity of alleles (different forms of genes) within a population.

Key Points:

- Mutations, genetic recombination, and gene flow introduce new genetic variations.
- Genetic variation is the raw material upon which natural selection acts.
- It leads to differences in traits among individuals, providing the basis for adaptation and evolution.

3. Genetic Drift:

- **Definition:** Genetic drift is the random fluctuation of allele frequencies in a population over generations due to chance events.

Key Points:

- Genetic drift is more prominent in small populations where chance events can have a significant impact.
- It can lead to the loss of genetic diversity and fixation of certain alleles, especially in small or isolated populations.

4. Gene Flow:

- **Definition:** Gene flow is the exchange of genetic material between different populations of the same species.

Key Points:

- Gene flow reduces genetic differences between populations.
- It introduces new alleles into a population, increasing genetic diversity.
- Gene flow can counteract the effects of genetic drift and natural selection, promoting genetic homogeneity among populations.

5. Mutation:

- **Definition:** Mutation is a random change in an organism's DNA sequence.

Key Points:

- Mutations are the ultimate source of genetic variation.
- Most mutations are neutral or harmful, but occasionally they create new beneficial traits.
- Beneficial mutations can be favored by natural selection, leading to adaptive evolution.

6. Sexual Selection:

- **Definition:** Sexual selection is a type of natural selection that favors traits that increase an organism's mating success.

Key Points:

- Sexual selection can lead to the evolution of elaborate traits in males (e.g., colorful plumage) or preferences in females for certain traits.
- These traits may not necessarily enhance survival but increase reproductive success, leading to their persistence in a population.

7. Adaptive Radiation:

- **Definition:** Adaptive radiation is the rapid diversification of a single ancestral lineage into a variety of forms to adapt to different environments.

Key Points:

- It often occurs in new or vacant ecological niches, promoting the evolution of diverse adaptations.
- Adaptive radiation results in the emergence of multiple species from a common ancestor, each adapted to a specific niche.

8. Co-Evolution:

- **Definition:** Co-evolution is the reciprocal evolutionary influence between two interacting species.

Key Points:

- Co-evolution can occur between predator and prey, host, and parasite, or mutualistic partners.
- Each species evolves in response to changes in the other, leading to a continuous cycle of adaptation and counter-adaptation.

In summary, the mechanisms of evolution—natural selection, genetic variation, genetic drift, gene flow, mutation, sexual selection, adaptive radiation, and co-evolution—interact in complex ways, driving the continuous change and diversification of life forms on Earth. These mechanisms are fundamental to our understanding of the biological world and its evolutionary history.

REFERENCES

1. Andreas, W. *The Origins of Evolutionary Innovations: A Theory of Transformative Change in Living Systems.*
2. Charles, W. F., & Jason, B. W. *Evolutionary Genetics: Concepts and Case Studies*
3. Eugene, V. K. *The Logic of Chance: The Nature and Origin of Biological Evolution.*
4. George, C. W. *Adaptation and Natural Selection.*
5. George, C. W. *Sex and Evolution.*
6. John, M. S., & Eörs, S. *The Major Transitions in Evolution.*
7. John, M. S., & Eörs, S. *The Origins of Life: From the Birth of Life to the Origin of Language.*
8. Martin, A. N. *Evolutionary Dynamics: Exploring the Equations of Life.*
9. Masatoshi, N. *Mutation-Driven Evolution.*
10. Massimo, P., & Gerd, B. M. *Evolution: The Extended Synthesis.*
11. Matt, R. *Genome: The Autobiography of a Species in 23 Chapters.*
12. Matt, R. *The Red Queen: Sex and the Evolution of Human Nature.*
13. Motoo, K. *The Neutral Theory of Molecular Evolution.*
14. Richard, L. B., & Eva, J. *Natural Selection and Social Theory: Selected Papers of Robert Trivers.*
15. Stuart, A. K. *The Origins of Order: Self-Organization and Selection in Evolution.*

CHAPTER 9

The Extended Evolutionary Synthesis

CONTENTS

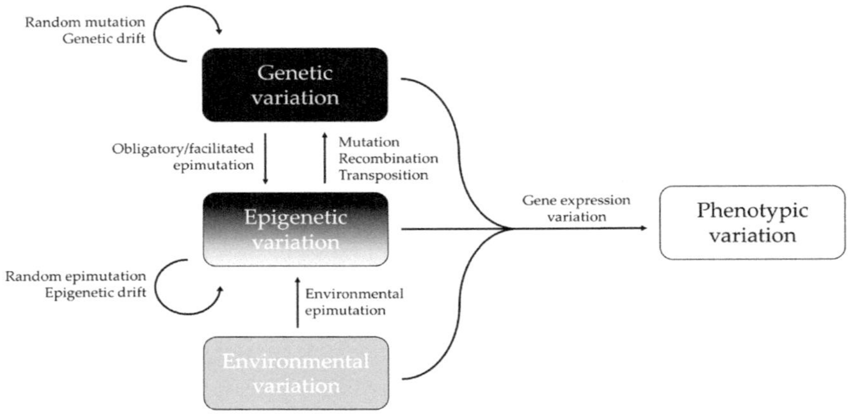

Figure 9.1. Extended evolutionary synthesis.

Source: Chapelle, V., & Silvestre, F., (2022). Population epigenetics: The extent of DNA methylation variation in wild animal populations. Epigenomes, 6(4), 31. https://doi.org/10.3390/epigenomes6040031; Creative Commons License.

9.1. BEYOND THE MODERN SYNTHESIS

The concept of "Beyond the Modern Synthesis" refers to developments and advancements in evolutionary biology that have occurred after the formulation of the Modern Synthesis, also known as the Neo-Darwinian synthesis. The Modern Synthesis, which emerged in the mid-20th century, integrated Mendelian genetics with Darwin's theory of natural selection, providing a comprehensive explanation for the mechanisms of evolution. However, since its inception, scientific understanding has continued to evolve, leading to new insights and perspectives in the field of evolutionary biology. Here are some key areas that represent the advancements made beyond the modern synthesis:

1. Genomic Evolution:

- **Genomic Studies:** Advances in genomics, including whole-genome sequencing and comparative genomics, have provided deep insights into the genetic basis of evolution. Comparative genomics allows scientists to compare entire genomes among different species, revealing patterns of conservation and variation that inform our understanding of evolutionary relationships.

2. Evo-Devo (Evolutionary Developmental Biology):

- **Developmental Genetics:** Evo-Devo explores the relationship between developmental processes and evolution. It investigates how changes in gene regulation and expression during development contribute to the evolution of morphological diversity. This field has highlighted the role of regulatory genes (such as homeobox genes) in shaping body plans and evolutionary novelties.

3. Epigenetics and Inheritance:

- **Epigenetic Inheritance:** Epigenetic modifications, which can influence gene expression without changing the underlying DNA sequence, have been recognized as important factors in evolution. Changes in epigenetic marks can be heritable and can affect traits within populations.

4. Microbial Evolution: Studying the evolution of microorganisms, including bacteria and viruses, has provided valuable insights into the mechanisms of rapid evolution, adaptation to changing environments, and the evolution of antibiotic resistance. Microbial evolution is crucial in areas such as medicine and agriculture.

5. Coevolutionary Dynamics:

- **Coevolutionary Studies:** Ongoing research in coevolution explores the complex interactions between species, including mutualistic relationships, predator-prey dynamics, and host-parasite interactions. Understanding these dynamics provides insights into the evolutionary arms race and the development of specialized traits in interacting species.

6. Cultural Evolution and Social Evolution:

- **Cultural Evolution:** The study of cultural evolution explores the transmission of cultural traits, beliefs, and practices across human populations. Cultural evolution can interact with biological evolution, influencing human behavior and social structures.
- **Social Evolution:** It examines the evolution of social behaviors and structures in animal societies, including cooperation, altruism, and social hierarchies. It explores how social behaviors are shaped by natural selection and social dynamics.

7. Evolutionary Ecology and Conservation Biology:

- **Evolutionary Ecology:** This investigates how ecological interactions and environmental pressures influence evolutionary processes. It examines adaptation to specific ecological niches, ecological speciation, and the role of biotic and abiotic factors in shaping evolutionary trajectories.

- **Conservation Biology:** Understanding evolutionary processes is essential in conservation biology. Conservation efforts often involve preserving genetic diversity within populations, managing endangered species, and addressing the impacts of environmental change on evolutionary processes.

Hence, the study of evolution has expanded significantly beyond the Modern Synthesis, encompassing diverse fields such as genomics, developmental biology, epigenetics, microbiology, cultural evolution, and ecology. These advancements continue to deepen our understanding of the mechanisms of evolution and its relevance to various scientific disciplines and real-world applications.

9.2. EPIGENETICS AND NON-DARWINIAN MECHANISMS

1. Epigenetic Inheritance:
   - **Transgenerational Epigenetic Inheritance:** Epigenetic modifications, such as DNA methylation and histone modifications, can be passed from one generation to the next. These modifications can influence gene expression and phenotypic traits, leading to heritable variations that are not encoded in the DNA sequence itself.
2. Phenotypic Plasticity:
   - **Environmental Responsiveness:** Epigenetic changes can influence an organism's ability to respond to environmental cues. Phenotypic plasticity allows organisms to exhibit different traits in response to varying environmental conditions. This plasticity enables rapid adaptation to changing environments without relying solely on genetic mutations.
3. Epigenetic Variation and Evolutionary Innovation:
   - **Epigenetic Diversity:** Within populations, individuals can exhibit epigenetic diversity that is not directly related to genetic variation. Epigenetic changes can lead to novel phenotypes, creating a pool of diverse traits within a population. Natural selection can act on this epigenetic variation, driving evolutionary innovation and adaptation.

4. Cultural and Social Evolution:
    - **Cultural Inheritance:** In many species, especially humans, cultural practices and knowledge are transmitted socially. Cultural evolution, facilitated by social learning and communication, allows for the rapid spread of behaviors and innovations within populations. These cultural practices can significantly influence human societies and in turn, impact evolutionary processes.
5. Epigenetic Regulation and Evolutionary Potential:
    - **Regulatory Evolution:** Epigenetic modifications play a role in gene regulation. Changes in epigenetic regulation can lead to the evolution of new gene expression patterns, influencing development and phenotype. This regulatory evolution can drive morphological and behavioral diversification within species.
6. Interspecies Epigenetic Interactions:
    - **Epigenetic Crosstalk:** Interactions between different species can influence epigenetic states. For example, in host-parasite interactions, parasites can manipulate host epigenetic machinery to evade immune responses. These interactions can have significant consequences for both host and parasite evolution.
7. Epigenetic Memory and Environmental Adaptation:
    - **Environmental Memory:** Epigenetic changes can create a cellular memory of past environmental exposures. This memory can influence an organism's response to similar environmental conditions in the future, providing a mechanism for adaptation to recurring environmental challenges.
8. Challenges and Considerations:
    - **Stability and Heritability:** The stability and heritability of epigenetic changes across generations are areas of active research. Understanding the persistence of epigenetic modifications and their long-term evolutionary impact is a complex and evolving field.

In summary, epigenetics and non-Darwinian mechanisms provide additional layers of complexity to evolutionary processes. They allow for the inheritance of acquired traits, rapid responses to environmental changes, and the potential for novel adaptations. Integrating these mechanisms with classical Darwinian evolution provides a more comprehensive understanding of the diverse pathways through which species evolve and adapt to their environments.

9.3. EVOLUTIONARY DEVELOPMENTAL BIOLOGY (EVO-DEVO)

Evolutionary developmental biology (Evo-Devo) is a field of biology that explores the relationship between development and evolution. It seeks to understand how changes in developmental processes have led to the evolution of diverse forms and traits in organisms over geological time scales. Evo-Devo integrates principles from developmental biology with evolutionary theory, shedding light on the underlying mechanisms that drive morphological diversity and evolutionary innovation. Here's a comprehensive overview of Evo-Devo:

1. Historical Background:

- **Pre-Evo-Devo Era:** Traditional evolutionary biology focused on genetic changes and natural selection as the main drivers of evolution. Morphological differences were primarily attributed to genetic mutations.
- **Emergence of Evo-Devo:** Evo-Devo gained prominence in the late 20th century and early 21st century, highlighting the role of gene regulation, embryonic development, and environmental interactions in shaping evolutionary trajectories.

2. Key Concepts and Approaches:

- **Conserved Regulatory Networks:** Evo-Devo studies have revealed that many genes and regulatory networks controlling development are conserved across diverse species. Changes in the regulation of these genes can lead to morphological differences.
- **Heterochrony:** Evo-Devo explores heterochrony, changes in the timing of developmental events. Alterations in the timing of gene expression can result in the evolution of different adult forms from similar embryonic stages.
- **Modularity:** Organisms are composed of modular developmental units. Changes in the interactions between these modules can lead to the evolution of new structures without disrupting the entire developmental process.

3. Examples and Case Studies:

- **Homeotic Genes:** Such as Hox genes, play a crucial role in determining the identity of body segments during development. Changes in the expression patterns of these genes have been linked to morphological diversity in animals.
- **Evolution of Limbs:** Studying the evolution of limbs, such as in tetrapods, has provided insights into the role of regulatory changes in the evolution of diverse limb morphologies, including wings, legs, and fins.
- **Evolution of Eyes:** Evo-Devo research on eye development has elucidated the genetic and developmental processes underlying the evolution of complex eye structures in different animal taxa.

4. Evo-Devo and Evolutionary Transitions:

- **Major Evolutionary Transitions:** Evo-Devo explores major evolutionary transitions, such as the evolution of multicellularity, segmentation, and the development of complex organ systems. Understanding the underlying developmental mechanisms provides insights into the origins of biological complexity.
- **Evolution of Novel Traits:** Evo-Devo investigates how novel traits, such as feathers in birds or the placenta in mammals, have evolved through modifications in developmental processes.

5. Evo-Devo and Evolutionary Questions:

- **Origins of Morphological Diversity:** Evo-Devo addresses the origins of morphological diversity and the genetic and developmental changes that lead to the evolution of novel structures.
- **Evolutionary Constraints:** By studying developmental constraints, Evo-Devo explores why certain morphological forms are more likely to evolve than others, considering both genetic and physical limitations.

6. Applications and Relevance:

- **Medical and Biotechnological Implications:** Understanding the molecular and developmental basis of evolutionary changes has applications in fields like medicine and biotechnology, informing research on human diseases and genetic therapies.
- **Conservation Biology:** Evo-Devo provides insights into the evolutionary potential of species, helping conservationists understand how organisms might adapt to changing environments.

So, evolutionary developmental biology (Evo-Devo) bridges the gap between developmental processes and evolutionary outcomes. By examining the genetic and molecular mechanisms underlying morphological diversity, Evo-Devo enriches our understanding of evolution and provides a deeper insight into the origins of biological complexity in the natural world.

9.4. OVERVIEW OF THE EXTENDED EVOLUTIONARY SYNTHESIS

The extended evolutionary synthesis (EES) represents an updated and expanded framework for understanding the mechanisms of evolution. It builds upon the foundation of classical Darwinian evolution while incorporating modern advances in genetics, developmental biology, ecology, and other fields. Here's an overview of the key aspects of the extended evolutionary synthesis:

1. **Incorporating Developmental Biology:** EES emphasizes the role of developmental processes in evolution. It considers how phenotypic variation arises during development and how these developmental processes are subject to natural selection and other evolutionary forces.
2. **Genetic and Epigenetic Inheritance:** EES acknowledges the importance of both genetic and epigenetic mechanisms in inheritance. Epigenetic modifications, such as DNA methylation and histone modifications, can influence gene expression and can be passed down to subsequent generations, contributing to heritable variation.
3. **Niche Construction and Ecological Inheritance:** Organisms don't just adapt to their environments; they actively modify and construct their environments. EES incorporates the concept of niche construction, where organisms modify their ecological niches, leading to feedback loops between organisms and their environments. These modifications can influence the selection pressures on future generations.
4. **Plasticity and Phenotypic Variation:** EES emphasizes phenotypic plasticity, the ability of a single genotype to produce different phenotypes in different environments. This plasticity allows organisms to cope with environmental changes and can influence the direction of evolution.

5. **Evolutionary Capacitance and Innovation:** EES introduces the concept of evolutionary capacitance, where certain traits or genetic variations are "hidden" within a population and can be rapidly expressed under specific conditions. This can lead to the sudden appearance of novel traits or innovations during times of environmental change.
6. **Symbiosis and Cooperation:** EES recognizes the importance of symbiotic relationships and cooperation between organisms in evolution. Cooperative interactions, such as mutualism, can lead to the evolution of interdependent traits and ecological specialization.
7. **Integrating Multiple Levels of Selection:** EES considers selection acting at various levels, including genes, individuals, populations, and communities. It acknowledges that evolution operates not only at the level of individual organisms but also at the level of groups and ecosystems.
8. **Cultural Evolution:** EES expands the scope of evolution to include cultural evolution, where human culture and social behaviors evolve through processes of learning, imitation, and innovation. Cultural evolution can influence genetic evolution and vice versa.

In fact, the Extended Evolutionary Synthesis provides a broader and more integrative perspective on the processes of evolution. By incorporating insights from diverse fields of biology, it offers a more comprehensive understanding of the mechanisms that drive evolutionary change and the complexity of biological systems. The EES framework continues to be a topic of active research and debate within the scientific community, shaping our understanding of evolution in the 21st century.

REFERENCES

1. Alessandro, M. *Towards a Theory of Development.*
2. Alessandro, M., & Giuseppe, F. *Evolving Pathways: Key Themes in Evolutionary Developmental Biology.*
3. Ames B. N., (1979). Identifying environmental chemicals causing mutations and cancer. *Science (New York, N.Y.), 204*(4393), 587–593. https://doi.org/10.1126/science.373122 (accessed on 27 October 2023).
4. Beadle, G., & Edward, T., (1941). Genetic control of biochemical reactions in neurospora. *PNAS, 27*(11), 499–506.
5. Benedikt, H., & Brian, K. H. *Epigenetics: Linking Genotype and Phenotype in Development and Evolution.*
6. Brian, K. H., & Wendy, M. O. *Evolutionary Developmental Biology.*
7. Daniel, J. F. *Evolving: The Human Effect and Why It Matters.*
8. Doolittle, F. W., (2000). Uprooting the tree of life. *Scientific American, 282*(2), 72–77.
9. Eric, S., & Harold, J. M. *The Origin and Nature of Life on Earth: The Emergence of the Fourth Geosphere.*
10. Jacob, F., (1977). *Evolution and Tinkering* (Vol. 196, No. 4295, pp. 1161–1166). Science, New Series.
11. Karl, J. N., & Stuart, A. N. *Evolutionary Biology: A Plant Perspective.*
12. Margulis, L., (1970). *Origin of Eukaryotic Cells.* Yale Univ. Press.
13. Margulis, L., (1967). On the origin of mitosing cells. *J. Theor. Bio., 14*(3), 255–274. (Sagan at the time).
14. Martin, A. N. *Evolutionary Dynamics: Exploring the Equations of Life.*
15. Massimo, P., & Gerd, B. M. *Extended Evolutionary Synthesis: A New Paradigm for Understanding Biology.*
16. McGowan, C., Roberta, F., Alice, W., & James, T., (1998). Evidence for interspecies gene transfer in the evolution of 2,4-dichlorophenoxyacetic acid degraders. *Appl. Environ. Microbiol., 64*(10), 4089–4092. http://www.ncbi.nlm.nih.gov/pmc/articles/PMC106609/ (accessed on 27 October 2023).
17. Muller, H. J., (1946). *For the Discovery of the Production of Mutations by Means of X-Ray Irradiation.* http://www.nobelprize.org/nobel_prizes/medicine/laureates/1946/muller.html?print=1 (accessed on 27 October 2023).

18. Newport, F., (2012). In: *U.S. 46% Hold Creationist View of Human Origins: Highly Religious Americans Most Likely to Believe in Creationism.* http://www.gallup.com/poll/155003/hold-creationist-view-human-origins.aspx (accessed on 27 October 2023).
19. Ohno, S., (1970). *Evolution by Gene Duplication.* Springer Verlag.
20. Olendzenski, L., & Peter, G. J., (2009). Evolution of genes and organisms: The Tree/web of life in light of horizontal gene transfer in natural genetic engineering and natural genome editing. *Ann. N.Y. Acad. Sci., 1178*, 137–145.
21. Samir, O., & Ken, B. *Evolution and the Levels of Selection.*
22. Shapiro, J., (2011). *Evolution: A View from the 21st Century*. FT Press Science.
23. Shapiro, J., (2012). *Huffington Post Blog.* http://www.huffingtonpost.com/james-a-shapiro/ (accessed on 27 October 2023).
24. Woese, C., (2004). A new biology for a new century. *Microbiol. Mol. Biol. Rev., 68*(2), 173–186.

CHAPTER 10

Evolutionary Applications

CONTENTS

10.1. EVOLUTIONARY MEDICINE

Evolutionary Medicine, also known as Darwinian Medicine or Evolutionary Health, is a field that applies principles of evolutionary biology to understand health and disease in humans. It explores how evolutionary processes have shaped human biology, including the genetic and physiological mechanisms that influence our susceptibility to diseases. Here's a comprehensive overview of evolutionary medicine (Figure 10.1):

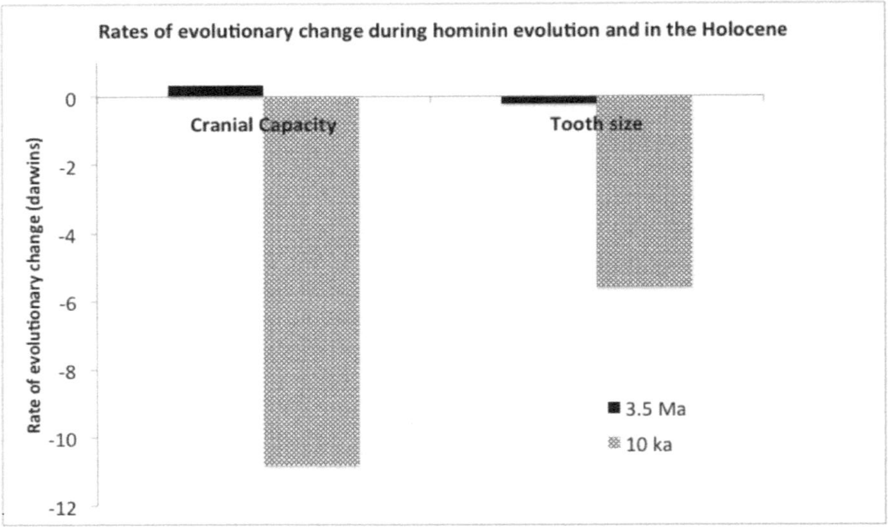

Figure 10.1. *Evolutionary applications.*

Source: Saniotis, A., & Henneberg, M., (2013). Evolutionary medicine and future of humanity: Will evolution have the final word? Humanities, 2(2), 278–291. https://doi.org/10.3390/h2020278; Creative Commons License.

1. Evolutionary Perspectives on Human Health:
   - **Mismatch Hypothesis:** Many modern diseases result from a mismatch between our evolved biological traits and our contemporary lifestyles. For example, genes adapted for a hunter-gatherer lifestyle can be maladaptive in the context of sedentary, high-calorie diets, leading to conditions like obesity and diabetes.
2. Evolutionary Insights into Diseases:
   - **Infectious Diseases:** Evolutionary perspectives help understand the co-evolutionary arms race between pathogens and the human immune system. This knowledge aids in predicting the emergence

of new infectious diseases and developing effective treatments.

- **Chronic Diseases:** Evolutionary medicine investigates the genetic basis of chronic diseases, including cardiovascular diseases and cancer, exploring how genetic variation influences disease susceptibility.

3. Genetic Variation and Disease Resistance:

- **Heterozygote Advantage:** Certain genetic variations that cause diseases in homozygotes may provide resistance to specific infections when present in heterozygotes. This phenomenon, such as the sickle cell trait's protection against malaria, illustrates the interplay between genetics and disease resistance.

4. Evolutionary Insights into Aging:

- **Evolutionary Theories of Aging:** Evolutionary Medicine explores theories of aging, such as antagonistic pleiotropy (genes with beneficial effects early in life and detrimental effects later) and the disposable soma theory (resources allocated between reproduction and maintenance). These theories shed light on the evolution of senescence.

5. Evolutionary Psychology and Mental Health:

- **Adaptations and Mental Health:** Evolutionary Psychology investigates mental health issues, exploring how certain traits or behaviors that were adaptive in ancestral environments might contribute to modern psychological disorders. For instance, anxiety might have evolved as a response to potential threats.
- **Mismatch and Mental Health:** Mismatches between ancestral environments and modern lifestyles, such as reduced physical activity and increased social stressors, can contribute to mental health issues. Evolutionary perspectives help identify these mismatches.

6. Evolutionary Perspectives in Treatment and Prevention:

- **Precision Medicine:** Understanding individual genetic variation from an evolutionary standpoint informs personalized medicine, tailoring treatments based on genetic profiles.
- **Preventive Strategies:** Evolutionary insights guide preventive strategies, emphasizing lifestyle changes aligned with our evolutionary heritage, including diet and physical activity patterns.

7. Cultural Evolution and Health Behaviors:
    - **Cultural Influences:** Cultural evolution has shaped human diets, social structures, and health behaviors. Evolutionary Medicine explores how cultural changes, such as the introduction of agriculture or processed foods, impact health outcomes.
8. Public Health and Policy Implications:
    - **Public Health Strategies:** Evolutionary perspectives inform public health policies, emphasizing the importance of promoting lifestyles in harmony with our evolutionary past to prevent diseases related to mismatched environments.
    - **Healthcare Planning:** Understanding genetic predispositions from an evolutionary context helps healthcare systems plan for the prevalence of certain diseases within populations.

In summary, Evolutionary Medicine provides valuable insights into the origins of human diseases, offering a holistic understanding of health and wellness. By considering our evolutionary history, this field enhances our ability to prevent, diagnose, and treat various health conditions, ultimately improving healthcare outcomes for individuals and populations.

10.2. ECOLOGICAL GENETICS

Ecological genetics is a field of biology that integrates principles from ecology and genetics to understand the genetic basis of adaptations to the environment, population dynamics, and evolutionary processes in natural populations. It explores how genetic variation within species is shaped by ecological factors and how this variation, in turn, influences the ecological interactions and evolutionary trajectories of populations. Here's a comprehensive overview of ecological genetics:

1. Genetic Variation in Natural Populations:
    - **Genetic Diversity:** Natural populations exhibit genetic variation, which includes differences in DNA sequences, gene expression patterns, and other heritable traits. Understanding this diversity is crucial for studying adaptation and evolution in changing environments.
2. Genetic Adaptations to the Environment:
    - **Local Adaptation:** Populations can evolve specific genetic adaptations to their local environments, such as tolerance to

temperature extremes, resistance to local diseases, or adaptations to specific food sources. These adaptations enhance the population's fitness in their ecological niche.

- **Gene-Environment Interactions:** Ecological genetics explores how genetic variants interact with environmental factors, influencing phenotypic traits. Such interactions are crucial in determining an organism's response to environmental changes.

3. Gene Flow and Genetic Structure:

- **Population Connectivity:** Gene flow, the movement of genetic material between populations, affects the genetic diversity and adaptive potential of populations. Ecological genetics investigates how barriers to gene flow, such as geographical features or ecological differences, influence genetic connectivity.
- **Genetic Structure:** Natural populations often exhibit genetic structuring, with distinct subpopulations adapted to different ecological conditions. Studying genetic structure provides insights into the evolutionary processes shaping populations.

4. Molecular Ecology:

- **DNA Markers:** Molecular techniques, including DNA markers like microsatellites and single nucleotide polymorphisms (SNPs), are used to assess genetic diversity, gene flow, and population structure. These markers aid in understanding the genetic basis of ecological traits.
- **Genomic Approaches:** Recent advances in genomics allow researchers to study the entire genome, revealing adaptive genes and genomic regions associated with ecological adaptations.

5. Quantitative Genetics and Ecological Traits:

- **Quantitative Trait Loci (QTL) Mapping:** Ecological geneticists use QTL mapping to identify genomic regions associated with complex ecological traits, such as body size, behavior, or reproductive strategies. Understanding the genetic basis of these traits provides insights into adaptation.
- **Heritability and Selection:** Estimating the heritability of ecological traits helps assess the potential for these traits to respond to natural selection. High heritability suggests that traits can evolve rapidly under selective pressures.

6. Evolutionary Dynamics in Changing Environments:

- **Climate Change and Evolution:** Ecological genetics explores how populations respond to climate change, studying adaptive evolution and potential genetic shifts in response to altered temperature, precipitation patterns, or habitat availability.
- **Human Impact:** Human activities, such as habitat destruction and pollution, affect natural populations. Ecological genetics helps assess the genetic impact of these disturbances and informs conservation efforts.

7. Conservation Genetics:

- **Genetic Diversity and Conservation:** Maintaining genetic diversity within populations is vital for their long-term survival. Conservation genetics assesses the genetic health of endangered species, guiding breeding programs and habitat restoration efforts to preserve genetic variability.
- **Population Viability Analysis:** Ecological genetics contributes to population viability analysis, assessing the genetic factors influencing population persistence and guiding conservation strategies.

So, ecological genetics provides a comprehensive understanding of the interplay between genetic variation, ecological interactions, and evolutionary processes in natural populations. By studying these dynamics, ecological geneticists contribute valuable insights to biodiversity conservation, evolutionary biology, and our understanding of how organisms adapt to their environments.

10.3. CONSERVATION BIOLOGY

Conservation biology is a multidisciplinary field that aims to understand, protect, and sustain Earth's biodiversity. It encompasses scientific research, management strategies, policy development, and public education to preserve the diversity of life on Earth. Here's a comprehensive overview of conservation biology:

1. Biodiversity and Its Importance:

- **Biodiversity Definition:** Biodiversity refers to the variety of life forms at all levels of biological organization, including genes, species, and ecosystems.

- **Importance of Biodiversity:** Biodiversity provides essential ecosystem services, such as clean air and water, pollination, climate regulation, and disease control. It also has intrinsic value, cultural significance, and economic importance.

2. Threats to Biodiversity:
   - **Habitat Loss:** Destruction, degradation, and fragmentation of natural habitats due to human activities like agriculture, urbanization, and logging.
   - **Climate Change:** Global warming and altered weather patterns threaten ecosystems and species, leading to shifts in habitats and endangering vulnerable species.
   - **Invasive Species:** Non-native species can outcompete native species, disrupt ecosystems, and lead to declines or extinctions.
   - **Pollution:** Pollution from various sources, including chemicals, plastics, and toxins, harms both aquatic and terrestrial ecosystems, affecting species survival.
   - **Overexploitation:** Unsustainable hunting, fishing, and harvesting of species for food, medicine, or trade can deplete populations and drive species to extinction.
3. Conservation Strategies:
   - **Protected Areas:** Establishing and maintaining protected areas, such as national parks and wildlife reserves, helps conserve biodiversity by preserving habitats and providing safe havens for endangered species.
   - **Habitat Restoration:** Efforts to restore degraded habitats, reforest areas, and rehabilitate ecosystems help reverse the impacts of habitat loss and degradation.
   - **Species Conservation:** Conservation programs focus on endangered species, including captive breeding, reintroduction, and genetic management to increase population sizes and genetic diversity.
   - **Community-based Conservation:** Involving local communities in conservation efforts ensures sustainable resource management, promotes coexistence, and supports livelihoods.
   - **Policy and Legislation:** Developing and enforcing environmental laws, international agreements (such as CITES), and conservation policies are essential for protecting biodiversity.

- **Conservation Genetics:** Genetic techniques help assess the genetic health of populations, identify unique genetic variants, and inform breeding programs to maintain genetic diversity.

4. Conservation Challenges and Ethical Considerations:
    - **Conservation Ethics:** Conservation biology grapples with ethical dilemmas, such as decisions about which species to save, reintroduction efforts, and the impact of conservation on indigenous communities.
    - **Human-Wildlife Conflict:** Finding solutions to conflicts between humans and wildlife, such as crop raiding by elephants or predation on livestock, is critical for both conservation and human well-being.
5. Conservation Education and Advocacy:
    - **Public Awareness:** Educating the public about biodiversity, its value, and conservation challenges raises awareness and fosters a sense of responsibility toward the environment.
    - **Advocacy and Policy Influence:** Conservation organizations advocate for policies that protect the environment, promote sustainable practices, and address the root causes of biodiversity loss.
6. Conservation Successes and Future Challenges:
    - **Success Stories:** Conservation efforts have led to the recovery of several endangered species, the establishment of protected areas, and the restoration of degraded ecosystems.
    - **Ongoing Challenges:** Continued habitat loss, climate change impacts, poaching, and the spread of invasive species pose ongoing challenges to conservation efforts.
    - Collaboration at global, national, and local levels is essential to address these challenges.

In summary, conservation biology plays a pivotal role in safeguarding Earth's biodiversity. By integrating scientific research, policy advocacy, community engagement, and public education, conservation biologists work toward a sustainable future where humans coexist harmoniously with the rich tapestry of life on our planet.

10.4. AGRICULTURAL AND INDUSTRIAL APPLICATIONS

Agricultural and industrial practices have significantly influenced the course of evolution, especially concerning domesticated plants, animals, and microorganisms. Here's how these applications have impacted evolution:

1. Domestication of Plants and Animals:
   - **Selective Breeding:** Humans have selectively bred plants and animals for desirable traits such as higher yield, resistance to diseases, and docility. This artificial selection has accelerated the evolution of domesticated species.
   - **Genetic Changes:** Selective breeding has led to genetic changes within populations, sometimes resulting in distinct breeds or varieties adapted to specific environments. This rapid evolution is a direct consequence of human intervention.
2. Crop Evolution and Agriculture:
   - **Agricultural Practices:** Such as irrigation, crop rotation, and soil management have influenced the evolution of crop plants. Crops that respond well to specific agricultural practices are favored, leading to adaptations within cultivated varieties.
   - **Genetic Modification:** Modern biotechnology allows scientists to introduce specific genes into plants, leading to genetically modified organisms (GMOs). This targeted genetic modification accelerates the evolution of traits like pest resistance or improved nutritional content.
3. Livestock Evolution:
   - **Animal Husbandry:** Domestication and selective breeding of livestock have led to the evolution of traits such as increased milk production, muscle mass, or wool quality. These changes are evolutionary responses to the preferences and needs of human populations.
   - **Industrial Farming:** Intensive farming practices, common in the industrial era, have led to rapid evolution in livestock populations. Animals are bred for specific traits to maximize productivity and economic efficiency.
4. Microbial Evolution and Industrial Applications:
   - **Industrial Fermentation:** Microorganisms like bacteria and yeast

have been harnessed for industrial processes such as fermentation for the production of beer, wine, and antibiotics. Over time, strains with desired fermentation traits have been selected, leading to the evolution of specialized industrial microorganisms.

- **Bioremediation:** Certain bacteria and fungi have evolved natural abilities to break down pollutants. In industrial applications, these organisms are used in bioremediation processes to clean up contaminated environments.

5. Pesticides and Resistance Evolution:

- **Pesticide Use:** The widespread use of pesticides in agriculture exerts selective pressure on pest populations. Pests with genetic variations that confer resistance to pesticides survive and reproduce, leading to the evolution of resistant pest populations.
- **Herbicide Resistance:** Similarly, the use of herbicides in agriculture has led to the evolution of herbicide-resistant weed species. These weeds have genetic mutations that allow them to survive exposure to herbicides.

6. Aquaculture and Evolutionary Impacts:

- **Fish Farming:** Aquaculture practices, including selective breeding, have led to the rapid evolution of fish populations. Fish are bred for traits like growth rate and disease resistance, resulting in genetically distinct farm-raised varieties.

In nutshell, agricultural and industrial applications have significantly influenced the evolution of various organisms, both intentionally through selective breeding and unintentionally through the unintended consequences of human practices. Understanding these evolutionary processes is crucial for sustainable agriculture, biodiversity conservation, and the responsible application of biotechnological advancements.

10.5. EVOLUTIONARY PSYCHOLOGY

Evolutionary psychology is a scientific discipline that explores how human psychological traits and behaviors have evolved over time through the process of natural selection. It seeks to understand the cognitive, emotional, and social aspects of the human mind by examining how they might have provided adaptive advantages to our ancestors. Here's a comprehensive overview of evolutionary psychology:

1. Evolutionary Foundations:
    - **Adaptation:** Evolutionary psychology posits that many human psychological traits and behaviors are adaptations shaped by natural selection. These adaptations helped our ancestors solve recurrent problems related to survival, reproduction, and social interaction.
    - **Environment of Evolutionary Adaptedness (EEA):** Evolutionary psychologists often refer to the EEA, which represents the ancestral environment in which human psychological traits evolved. Understanding the EEA helps in analyzing the adaptive significance of various psychological phenomena.
2. Key Concepts and Principles:
    - **Modularity:** Evolutionary psychologists propose that the human mind consists of specialized modules or mental mechanisms, each designed to solve specific adaptive problems. These modules contribute to various cognitive functions, such as language acquisition, social reasoning, and mate selection.
    - **Sexual Selection:** Evolutionary psychology investigates how mate preferences and mating behaviors have evolved through sexual selection, leading to the development of specific traits and preferences that enhance reproductive success.
    - **Parental Investment Theory:** This theory explains differences in mating strategies and behaviors based on the unequal parental investment between males and females. It predicts gender-specific mating preferences and behaviors.
3. Evolutionary Explanations for Human Behavior:
    - **Altruism and Cooperation:** Evolutionary psychology explores the origins of altruism and cooperation, suggesting that these behaviors can enhance inclusive fitness, where individuals help others, even at a cost to themselves, if it benefits their genetic relatives' survival and reproduction.
    - **Aggression and Competition:** Evolutionary perspectives on aggression propose that aggressive behaviors might have evolved as strategies to compete for resources, mates, or social status. Aggression can be seen as a means to resolve conflicts and establish dominance hierarchies.

4. Cognitive Evolution:

- **Language Acquisition:** Evolutionary psychologists investigate the cognitive mechanisms underlying language acquisition. The ability to acquire language is considered an adaptation that facilitated communication and social cooperation among early humans.
- **Theory of Mind:** The development of theory of mind, the ability to attribute mental states to oneself and others, is explored from an evolutionary perspective, highlighting its adaptive significance in social interactions and cooperation.

5. Evolutionary Psychology and Modern Behaviors:

- **Mate Preferences:** Evolutionary psychology examines modern mate preferences, including factors such as physical attractiveness, social status, and resources. These preferences are often viewed through the lens of evolutionary strategies for maximizing reproductive success.
- **Parenting Behaviors:** Evolutionary psychology studies parenting behaviors, exploring how parental investment, parental care, and parent-child bonding have evolved to ensure the survival and well-being of offspring.

6. Critiques and Controversies:

- **Sociocultural Influences:** Critics argue that evolutionary psychology often overlooks the role of sociocultural factors in shaping human behavior. They emphasize the importance of social and cultural contexts in understanding psychological phenomena.
- **Ethical Considerations:** Some ethical concerns arise from the potential misuse of evolutionary explanations for justifying gender roles, stereotypes, or discriminatory practices. Ethical guidelines are important in evolutionary psychology research.

In summary, evolutionary psychology provides valuable insights into the origins of human behaviors and cognition. By examining psychological phenomena through an evolutionary lens, researchers gain a deeper understanding of the adaptive functions of human traits and behaviors, shedding light on the complex interplay between biology, environment, and culture in shaping the human mind.

10.6. GROUP SELECTION

Group selection is a controversial evolutionary theory proposing that natural selection operates not only at the level of individuals but also at the level of entire groups or populations. It suggests that traits promoting the welfare of a group can be favored by selection, even if they may not offer immediate advantages to individual members. Here's a comprehensive overview of group selection in evolution:

1. Basic Concepts:

- **Individual Selection:** Classical Darwinian evolution emphasizes individual selection, where traits enhancing an individual's reproductive success are favored by natural selection. This process drives the evolution of traits that benefit individuals, even at the expense of the group.
- **Group Selection:** It posits that traits promoting cooperation, altruism, or social behaviors can be favored if they enhance the overall survival and reproductive success of a group. This perspective focuses on the fitness of groups of individuals rather than individual fitness alone.

2. Examples of Altruistic Behavior:

- **Altruism in Nature:** Altruistic behaviors, where individuals help others at a cost to themselves, are observed in various species. Examples include alarm calls in meerkats and cooperative hunting in wolves. These behaviors seem altruistic at the individual level but can benefit the group as a whole.

3. Conditions Favoring Group Selection:

- **High Relatedness:** Group selection is more likely to operate in populations where individuals are closely related. If individuals within a group share a significant proportion of their genes, altruistic traits directed at relatives (kin selection) can enhance the inclusive fitness of the group.
- **Between-Group Competition:** Intense competition between groups can favor traits that promote cooperation and cohesion within groups. In such scenarios, groups with cooperative individuals may outcompete other groups, leading to the spread of group-beneficial traits.

4. Controversies and Criticisms:

- **Levels of Selection:** The debate over levels of selection is central to the controversy surrounding group selection. Critics argue that selection acting at the level of individuals or genes can often explain apparently altruistic behaviors, without invoking group selection.
- **Free Riders:** The presence of "free riders" (individuals benefiting from cooperation without contributing) poses a challenge to the stability of altruistic traits within groups. Mechanisms like punishment of non-cooperators or preferential interactions with cooperators might mitigate free rider problems.

5. Multilevel Selection Theory:

- **Multilevel Selection:** This theory integrates individual selection, group selection, and kin selection into a unified framework. It recognizes that different levels of selection can operate simultaneously, and that the relative importance of each level depends on the specific context and the traits under consideration.

6. Human Societies and Culture:

- **Cultural Evolution:** Some proponents of group selection argue that cultural evolution, including the development of moral norms and cooperative behaviors, is driven by group-level selection. Cultural traits promoting cooperation and social cohesion may enhance the success of human groups.

In short, the concept of group selection remains a topic of debate in evolutionary biology. While it has found support in certain contexts, the interactions between different levels of selection and the complexities of social behaviors continue to be subjects of research and discussion. Multilevel selection theories provide a nuanced framework for understanding the interplay between individual and group-level adaptations in the evolution of social traits.

10.6.1. Cultural Evolution

Cultural evolution, within the context of evolutionary theory, represents the adaptive changes in human societies' beliefs, customs, technologies, and social structures over time. Integrating cultural evolution into evolutionary theory provides a nuanced understanding of how human culture influences and is influenced by biological evolution. Here's a comprehensive overview of cultural evolution within the framework of evolutionary theory:

1. Cultural Transmission and Evolution:
    - **Cultural Replication:** Cultural traits, ranging from languages to tools, are transmitted through generations via social learning, mirroring the principles of genetic replication in biological evolution.
    - **Cumulative Cultural Evolution:** Unique to humans, cultural evolution allows for the accumulation of knowledge and innovations over generations, leading to increasingly sophisticated technologies and social systems.
2. Gene-Culture Coevolution:
    - **Interaction Between Genes and Culture:** Human genetic evolution and cultural evolution are intertwined. Genetic predispositions influence cultural learning, and cultural practices, in turn, affect human selection pressures and adaptation.
    - **Example:** Lactose Tolerance: The coevolution of genetic lactose tolerance and dairy farming practices illustrates how cultural shifts can drive genetic adaptations in human populations.
3. Cultural Evolution and Social Behavior:
    - **Norms and Cooperation:** Cultural norms and moral values guide human behavior, fostering cooperation and societal harmony. Norms regulating cooperation and punishing selfish behavior enhance social cohesion.
    - **Reciprocity and Altruism:** Cultural practices promoting reciprocity and altruism, such as gift-giving and cooperation in hunter-gatherer societies, illustrate the evolutionary advantages of prosocial behaviors.
4. Cultural Evolution and Language:
    - **Origin of Language:** Cultural evolution theories explore the emergence of language as a cultural adaptation, enabling efficient communication and knowledge transfer among early human communities.
    - **Diversity of Languages:** The diversity of languages worldwide reflects cultural evolution, with languages adapting and diversifying based on historical migrations, interactions, and environmental influences.

5. Cultural Evolution and Technology:
   - **Technological Advancements:** Cultural evolution drives the development of tools and technologies, enhancing human survival and resource utilization. Advancements, from the Stone Age to the digital era, demonstrate cultural adaptation to changing environments.
   - **Innovation and Diffusion:** Cultural innovation, such as the invention of agriculture, spreads through diffusion, influencing neighboring cultures and leading to complex societal transformations.
6. Cultural Evolution in Modern Contexts:
   - **Cultural Evolution in Globalization:** Globalization accelerates cultural exchange, leading to the blending of traditions, languages, and practices. The adaptation of cultures to global challenges illustrates the ongoing process of cultural evolution.
   - **Cultural Preservation:** Cultural evolution also involves efforts to preserve indigenous knowledge and traditions threatened by globalization, emphasizing the importance of cultural diversity in the modern world.
7. Ethical Implications and Cultural Evolution:
   - **Ethical Considerations:** Understanding cultural evolution raises ethical questions about cultural relativism, the preservation of cultural practices, and the role of cultural diversity in promoting global understanding and tolerance.
   - **Cultural Ethics:** Cultural evolution theories inform discussions about cultural ethics, emphasizing the need for respectful engagement with diverse cultures and the recognition of the evolutionary significance of cultural practices.

Integrating cultural evolution into evolutionary theory provides a holistic perspective on human adaptation and societal development. Cultural evolution illustrates the dynamic interplay between biological and cultural factors, emphasizing the role of culture as a crucial adaptive mechanism in the human evolutionary story. This comprehensive understanding enriches our insights into the complex and multifaceted nature of human evolution.

10.6.2. Cultural Group Selection

Cultural group selection is a theoretical framework within evolutionary theory that posits the selection of cultural traits at the group level. Unlike individual selection, which focuses on the reproductive success of individuals, cultural group selection suggests that certain cultural traits can enhance the survival and reproductive success of entire groups or societies. Here's a comprehensive overview of cultural group selection:

1. Group-Level Traits:
    - **Definition:** Cultural group selection refers to the process by which cultural traits that promote group cooperation, social cohesion, and collective well-being are favored over time, leading to the success and proliferation of entire societies.
    - **Examples of Group-Level Traits:** Altruism, cooperation, reciprocity, cultural norms, and institutions that foster social harmony are examples of group-level traits that can be favored by cultural group selection.
2. Mechanisms and Conditions:
    - **Inter-Group Competition:** Cultural group selection operates effectively in scenarios where different groups compete for resources, territory, or survival. Groups with cooperative and altruistic members are more likely to outcompete other groups, leading to the selection of group-beneficial traits.
    - **Cultural Variation:** Variation in cultural practices between groups is essential for cultural group selection to occur. When different groups exhibit diverse cultural traits, competition, and selection can lead to the spread of advantageous traits.
3. Altruism and Cooperation:
    - **Evolution of Altruistic Behaviors:** Cultural group selection offers explanations for the evolution of altruistic behaviors within societies. Altruistic individuals who contribute to the well-being of their group, even at a personal cost, can enhance the overall success of the group, leading to the propagation of altruistic tendencies.
    - **Cooperation and Social Norms:** Social norms and cultural practices that promote cooperation and punish selfish behavior are conducive to cultural group selection. Such norms facilitate collaboration within groups, fostering unity and collective action.
4. Cultural Group Selection in Human Societies:
    - **Historical Examples:** Historical examples, such as cooperative

hunting practices in early human societies or communal agricultural systems, exemplify cultural group selection. These practices enhanced group survival and were likely favored by selection.

- **Cultural Evolution and Complex Societies:** Cultural group selection theories are used to understand the emergence and maintenance of complex human societies. Institutions, governance structures, and cultural norms that promote cooperation have been crucial for the success of large-scale civilizations.

5. Critiques and Controversies:

- **Debate on Levels of Selection:** Cultural group selection is a topic of debate in evolutionary theory. Critics argue that individual-level selection and cultural transmission can explain cooperative behaviors without invoking group-level selection.
- **Role of Cultural Transmission:** The mechanisms of cultural transmission, such as conformity and social learning, also play a significant role in the spread of cultural traits. Cultural group selection theories must consider the interplay between these transmission mechanisms and group-level selection pressures.

6. Modern Applications and Implications:

- **Contemporary Challenges:** Cultural group selection theories are applied to modern challenges, such as climate change and global cooperation. Understanding how cultural traits promoting cooperation and environmental stewardship can be selected at the group level is crucial for addressing these challenges.
- **Ethical Considerations:** Cultural group selection raises ethical questions about cultural diversity, human rights, and the preservation of indigenous knowledge. Balancing the importance of cultural diversity with universal ethical principles is a complex aspect of this theory.

In a nutshell, cultural group selection provides insights into the evolution of cooperation, altruism, and social norms within human societies. While the theory is debated, its exploration offers valuable perspectives on the role of cultural traits in shaping the success and resilience of human communities across history and in the face of contemporary challenges.

10.6.3. Dual Inheritance Theory

Dual Inheritance Theory (DIT) is an interdisciplinary framework that combines evolutionary biology and cultural evolution to understand the complex interactions between genetic and cultural evolution. It proposes that human evolution is shaped not only by genetic inheritance but also by the transmission of cultural knowledge and practices. Here's a comprehensive overview of Dual Inheritance Theory:

1. Genetic Evolution:

- **Selection and Adaptation:** DIT acknowledges the role of genetic evolution through natural selection. Genetic traits that enhance survival and reproduction are passed down to subsequent generations, contributing to the biological adaptation of populations to their environments.
- **Evolutionary Psychology:** DIT incorporates insights from evolutionary psychology, emphasizing the role of evolved cognitive mechanisms that influence cultural learning and social behaviors.

2. Cultural Evolution:

- **Cultural Transmission:** DIT highlights cultural transmission as a parallel process to genetic inheritance. Cultural knowledge, beliefs, and practices are transmitted through social learning, allowing the accumulation of cultural traits across generations.
- **Cultural Adaptation:** Cultural evolution enables populations to adapt rapidly to changing environments. Cultural innovations, such as tools, agricultural practices, and social norms, spread within and between populations, shaping human societies.

3. Gene-Culture Coevolution:

- **Interactive Evolution:** DIT explores the reciprocal interactions between genetic and cultural evolution. Cultural practices can influence selection pressures on genes, while genetic predispositions can shape cultural behaviors. This coevolutionary process leads to the dynamic adaptation of both genetic and cultural traits.
- **Example:** Lactose Tolerance: The coevolution of lactose tolerance and dairy farming practices illustrates the interplay between genetic adaptation (lactose tolerance) and cultural innovation (domestication of animals and milk consumption).

4. Transmission and Innovation:

- **Modes of Cultural Transmission:** DIT distinguishes between vertical transmission (from parents to offspring) and oblique transmission (from other relatives or non-kin). Additionally, horizontal transmission (between unrelated individuals) and oblique transmission contribute to the spread of cultural traits.
- **Cultural Innovation:** These arise from individual creativity or modification of existing practices. Innovations can spread rapidly within a population, leading to cultural diversification and adaptation.

5. Human Cooperation and Social Norms:

- **Evolution of Cooperation:** DIT explores the evolution of cooperation and altruism within social groups. Cultural norms and institutions that promote cooperation and punish free riders enhance social cohesion and cooperation, contributing to the success of human societies.
- **Punishment and Norm Enforcement:** DIT emphasizes the role of punishment mechanisms in maintaining social norms. Individuals may punish norm violators, even at a personal cost, to preserve the integrity of cooperative behaviors.

6. Critiques and Debates:

- **Levels of Selection:** DIT raises questions about the levels of selection, encompassing genes, individuals, groups, and cultures. Debates arise regarding the relative importance of these levels in shaping human evolution and behavior.
- **Complexity and Interactions:** The interactions between genes and culture are complex and multifaceted. Critics argue that quantifying these interactions and understanding their causal relationships pose significant challenges.

7. Applications and Future Directions:

- **Cultural Conservation:** DIT informs efforts to conserve endangered cultures and indigenous knowledge. Understanding the dynamics of cultural transmission aids in the preservation of cultural diversity.
- **Policy Implications:** DIT has implications for various fields, including education, public health, and social policy. Cultural interventions that align with human predispositions and social

learning mechanisms can be more effective in inducing behavioral change.

In fact, the Dual Inheritance Theory provides a holistic framework for understanding human evolution, incorporating the dynamic interplay between genetic and cultural evolution. By recognizing the significance of cultural transmission, DIT enriches our understanding of human behaviors, cooperation, and societal adaptation, offering valuable insights into the complexities of human evolution and culture.

10.7. HOLOGENOME THEORY OF EVOLUTION

The hologenome theory of evolution (HTE) is a modern evolutionary framework that extends the classical concept of evolution by incorporating the symbiotic relationship between hosts (such as organisms) and their associated microorganisms, including bacteria, viruses, and other microorganisms. According to this theory, the host, and its symbiotic microorganisms, collectively referred to as the hologenome, evolve as a single unit. Here's a comprehensive overview of the Hologenome Theory of Evolution:

1. The Hologenome Concept:
   - **Definition:** The hologenome represents the collective genome of the host organism and its symbiotic microorganisms. It includes the host's nuclear genome, mitochondrial genome, and the genomes of all associated microorganisms.
   - **Symbiosis and Evolution:** HTE posits that the host and its symbiotic microorganisms are intricately connected in a mutually beneficial relationship. Changes in the hologenome, driven by symbiotic interactions, can lead to evolutionary adaptations in the host population.
2. Holobionts and Hologenomes:
   - **Holobionts:** It is an organism and its associated symbiotic microorganisms treated as a single evolutionary unit. Holobionts can be plants, animals, or other multicellular organisms.
   - **Hologenomic Variability:** The hologenome is dynamic and can change over time due to various factors, including symbiotic associations, environmental influences, and horizontal gene transfer between microorganisms.

3. Transmission and Selection:

- **Vertical and Horizontal Transmission:** Symbiotic microorganisms can be transmitted vertically from parents to offspring or horizontally between individuals. Both types of transmission influence the composition and evolution of the hologenome.
- **Selection Pressures:** Selection acts on both the host and its symbiotic microorganisms. Beneficial microorganisms that enhance the host's fitness are more likely to be retained in the hologenome, leading to adaptive evolution.

4. Holobiont Evolutionary Processes:

- **Cooperation and Conflict:** HTE explores the balance between cooperation and conflict among the host and its microorganisms. Cooperative interactions, such as nutrient exchange, can enhance holobiont fitness, while conflicts may arise due to resource competition.
- **Role of Microbiota:** The microbiota plays a crucial role in host immune system development, nutrient metabolism, and defense against pathogens. Changes in the microbiota composition can influence the host's evolutionary trajectory.

5. Holobiont Adaptation and Health:

- **Host-Microbiota Interactions:** The symbiotic microorganisms contribute to the host's adaptation to diverse environments, including extreme conditions. The microbiota's role in host health and disease susceptibility is a key area of research in HTE.
- **Holobiont Resilience:** HTE emphasizes the holobiont's ability to adapt and evolve in response to changing environments. A diverse and stable microbiota can enhance the holobiont's resilience to environmental stressors.

6. Applications and Future Directions:

- **Eco-Evolutionary Dynamics:** HTE has implications for understanding eco-evolutionary processes in natural ecosystems. Holobionts, including plants and animals, influence ecological interactions and ecosystem functioning.
- **Biomedical and Agricultural Applications:** HTE research informs medical and agricultural practices. Understanding the host-microbiota interactions can lead to innovative approaches in

disease prevention, probiotics, and sustainable agriculture.

In summary, the Hologenome Theory of Evolution provides a novel perspective on evolutionary processes, emphasizing the interconnectedness of hosts and their symbiotic microorganisms. By considering holobionts as evolutionary units, HTE expands our understanding of adaptation, diversity, and resilience in the natural world. This theory continues to inspire research across various disciplines, shaping our knowledge of evolutionary biology and its applications in diverse fields.

10.8. MISSING HERITABILITY PROBLEM

The Missing Heritability Problem refers to the gap between the heritability of traits as estimated from family, twin, and adoption studies and the portion of heritability explained by identified genetic variants. This phenomenon challenges our understanding of the genetic basis of complex traits and has implications for evolutionary theory. Here's a comprehensive overview of the Missing Heritability Problem in the context of evolutionary theory:

1. Heritability Estimates:

- **Definition:** Heritability measures the proportion of phenotypic variation in a population that is attributed to genetic factors.
- **Empirical Studies:** Traditional studies estimate heritability by comparing traits among family members. These studies suggest a significant genetic component for various traits, including height, intelligence, and susceptibility to diseases.

2. Genome-Wide Association Studies (GWAS):

- **Identification of Genetic Variants:** GWAS reports analyze thousands of genomes to identify genetic markers associated with specific traits or diseases.
- **Limited Explained Variation:** Despite GWAS identifying numerous genetic variants, these variants collectively explain only a small fraction of the heritability observed in family studies, leading to the Missing Heritability Problem.

3. Factors Contributing to Missing Heritability:

- **Rare Variants:** Rare genetic variants with large effects on traits are often missed by GWAS due to their low frequency in the population.
- **Gene-Gene Interactions:** Epistatic interactions, where multiple

genes influence a trait together, are challenging to detect but are crucial for understanding heritability.

- **Gene-Environment Interactions:** Environmental factors can modify the effects of genetic variants, complicating the assessment of genetic heritability.

4. Implications for Evolutionary Theory:

- **Adaptation and Evolution:** Understanding the genetic basis of traits is fundamental for evolutionary biology. Traits under selection pressure are expected to have a substantial genetic component.
- **Genetic Diversity:** The missing heritability problem raises questions about the sources of genetic diversity. Unidentified genetic variants might contribute significantly to the genetic variation within populations.
- **Evolutionary Potential:** Identifying missing heritability is crucial for understanding a population's potential to adapt to changing environments over evolutionary time scales.

5. Emerging Research Avenues:

- **Functional Genomics:** Studying gene function through functional genomics approaches, like CRISPR/Cas9 gene editing, helps researchers understand the roles of specific genes in trait variation.
- **Non-Coding Regions:** Investigating non-coding regions of the genome, including regulatory elements, can provide insights into gene expression regulation and phenotypic variation.
- **Integration of Multi-Omics Data:** Integrating genomic, epigenomic, transcriptomic, and proteomic data can provide a comprehensive view of the genetic basis of traits, potentially uncovering missing heritability.

6. Challenges and Future Directions:

- **Statistical Methods:** Developing advanced statistical methods capable of detecting complex gene-gene and gene-environment interactions is essential for uncovering missing heritability.
- **Longitudinal Studies:** Long-term studies tracking individuals and families over generations are invaluable for understanding the heritability patterns across diverse traits.

- **Collaborative Research:** Collaborations between geneticists, statisticians, and evolutionary biologists are vital for developing interdisciplinary approaches to address the Missing Heritability Problem.

Thus, the missing heritability problem challenges our understanding of the genetic basis of traits, including those crucial for evolutionary processes. Addressing this problem not only enhances our knowledge of heritability but also deepens our comprehension of evolution, adaptation, and the complexities of genetic variation within populations. Continued research efforts, incorporating diverse methodologies and interdisciplinary collaborations, are essential to unraveling the mysteries of missing heritability in the context of evolutionary theory.

10.9. MOLECULAR EVOLUTION

Molecular evolution is the study of how genes and proteins evolve at the molecular level, exploring the mechanisms and processes that lead to genetic diversity and evolutionary change within populations and species. Here's a comprehensive overview of molecular evolution:

1. Genetic Variation:

- **Mutation:** Genetic variation arises from mutations, which are changes in the DNA sequence. Point mutations, insertions, deletions, and duplications can occur, leading to different alleles in populations.
- **Genetic Recombination:** During meiosis, genetic material is exchanged between homologous chromosomes, leading to new combinations of alleles in offspring.

2. Genetic Drift:

- **Definition:** Genetic drift refers to the random fluctuations of allele frequencies in small populations over generations. It can lead to the fixation or loss of alleles by chance, especially in small or isolated populations.
- **Founder Effect and Population Bottlenecks:** Founder effect occurs when a small group establishes a new population, potentially carrying a subset of the genetic variation of the original population. Population bottlenecks occur when a population is drastically reduced, leading to reduced genetic diversity.

3. Natural Selection:

- **Positive Selection:** It increases the frequency of advantageous alleles, enhancing the adaptation of populations to their environment.
- **Negative Selection:** It eliminates deleterious alleles, preserving the functional integrity of genes and proteins.
- **Balancing Selection:** It maintains multiple alleles in a population, which can be advantageous under different environmental conditions.

4. Genomic Evolution:

- **Gene Duplication:** It events create paralogous genes, which can evolve new functions or specialize in different tasks.
- **Horizontal Gene Transfer:** Transfer of genetic material between unrelated species can lead to the acquisition of new genes and traits, especially in prokaryotes.
- **Genome Size Evolution:** Genome sizes vary widely among organisms. Processes like gene duplication and the accumulation of non-coding DNA contribute to genome size evolution.

5. Molecular Clocks and Evolutionary Time Scales:

- **Molecular Clock Hypothesis:** Molecular clocks use the rate of genetic changes to estimate divergence times between species. Certain genes or genomic regions evolve at relatively constant rates and can serve as molecular clocks.
- **Calibrating Molecular Clocks:** Fossil records and geological events are used to calibrate molecular clocks, providing estimates of evolutionary events.

6. Evolutionary Genetics of Proteins:

- **Protein Structure and Function:** Changes in amino acid sequences can affect protein structure and function. Studying protein evolution helps us understand the relationship between genotype and phenotype.
- **Protein Families and Domains:** Proteins are classified into families and domains based on evolutionary relationships. Homologous proteins share a common ancestor and can provide insights into evolutionary histories.

7. Evolutionary Genomics and Comparative Genomics:

- **Comparative Genomics:** It involves comparing genomes across different species to identify similarities, differences, and evolutionary patterns.
- **Evolutionary Genomic Approaches:** Genomic studies enable the investigation of genome-wide evolutionary processes, including the identification of conserved elements, non-coding RNAs, and regulatory regions.

8. Applications and Future Directions:

- **Phylogenetics:** Molecular data are used to reconstruct evolutionary relationships among species, leading to the development of phylogenetic trees that depict the evolutionary history of life forms.
- **Medical and Biotechnological Applications:** Understanding molecular evolution is crucial in fields like medicine and biotechnology, where knowledge of genetic variation and adaptation can inform disease research and biotechnological innovations.

In summary, molecular evolution provides a powerful lens through which scientists can explore the intricate mechanisms of genetic change and adaptation over evolutionary time scales. This field continues to advance our understanding of the fundamental processes shaping the diversity of life on Earth and holds promise for various applications in science, medicine, and technology.

10.10. ASTROBIOLOGY

Astrobiology is a multidisciplinary scientific field that explores the potential for life beyond Earth, including its origins, evolution, and existence in the universe. Here's a comprehensive overview of astrobiology:

1. Origins of Life:

- **Abiogenesis:** Astrobiology investigates how life might have originated from non-living matter on Earth and potentially on other celestial bodies.
- **Early Earth Conditions:** Studying extreme environments on Earth, such as hydrothermal vents and acidic lakes, provides insights into the conditions that might have supported early life.

2. Life in Extreme Environments:

- **Extremophiles:** Astrobiologists study extremophiles, organisms thriving in extreme environments like deep ocean vents and acidic hot springs. These organisms provide clues about the limits of life and the potential habitability of other planets.

3. Search for Extraterrestrial Life:

- **Exoplanets:** Astronomers search for exoplanets within the habitable zone, where conditions might allow liquid water to exist. The discovery of exoplanets has expanded the search for extraterrestrial life beyond our solar system.
- **Mars Exploration:** Mars is a primary target in the search for past or present life. Rovers and landers have been sent to explore Martian geology and search for signs of microbial life.
- **Europa and Enceladus:** Moons like Europa (around Jupiter) and Enceladus (around Saturn) have subsurface oceans, raising the possibility of microbial life beneath their icy surfaces.
- **SETI:** The Search for Extraterrestrial Intelligence (SETI) uses radio telescopes to listen for signals from extraterrestrial civilizations.

4. Chemical Evolution and Astrochemistry:

- **Prebiotic Chemistry:** Astrobiologists study the chemical evolution of complex organic molecules, shedding light on the potential building blocks of life.
- **Astrochemistry:** This field explores the chemical composition of space, including the formation of organic molecules in interstellar clouds and their delivery to planets.

5. The Drake Equation and Fermi Paradox:

- **Drake Equation:** The Drake Equation estimates the potential number of extraterrestrial civilizations in our galaxy. It considers factors like star formation rates, planetary systems, and the likelihood of life evolving.
- **Fermi Paradox:** Despite the vast number of potentially habitable planets, humanity has not yet detected any extraterrestrial civilizations, raising the question of why we haven't encountered alien life.

6. Planetary Protection and Ethics:

- **Planetary Protection:** Measures are taken to prevent contamination of other celestial bodies by Earth microorganisms, ensuring that scientific exploration does not interfere with potential alien ecosystems.
- **Ethical Considerations:** Astrobiology raises ethical questions about the potential discovery of extraterrestrial life, including the implications for religion, society, and environmental ethics.

7. Astrobiology and the Future:

- **Space Missions:** Ongoing and future space missions, such as the Mars Sample Return mission, aim to bring back samples from other planets, offering new opportunities to study potential signs of life.
- **Technological Advancements:** Advancements in telescopes, space probes, and laboratory techniques continue to enhance our ability to search for extraterrestrial life and understand the conditions necessary for life's existence.

In summary, astrobiology explores the profound questions about the origins of life, the potential for life beyond Earth, and humanity's place in the universe. This interdisciplinary field bridges biology, astronomy, chemistry, and geology, offering intriguing possibilities for the future of scientific discovery and our understanding of life's existence in the cosmos.

10.11. PHYLOGENETICS

Phylogenetics is the study of evolutionary relationships among organisms, both living and extinct. It involves the analysis of genetic, morphological, and behavioral traits to understand the evolutionary history and diversification of life on Earth. Here's a comprehensive overview of phylogenetics:

1. Phylogenetic Trees:

- **Definition:** Phylogenetic trees depict the evolutionary relationships among species, illustrating their common ancestry and divergence over time.
- **Nodes and Branches:** Nodes represent common ancestors, and branches represent evolutionary lineages. The length of branches can indicate the amount of evolutionary change.
- **Rooted vs. Unrooted Trees:** Rooted trees have a specified

common ancestor, while unrooted trees depict relationships without specifying an ancestral point.

2. Molecular Phylogenetics:
   - **DNA Sequences:** Molecular phylogenetics analyzes DNA, RNA, or protein sequences to infer evolutionary relationships. DNA sequencing techniques have revolutionized the field, enabling high-resolution analyses.
   - **Phylogenetic Markers:** Specific genes or regions of the genome, such as ribosomal RNA or mitochondrial DNA, serve as molecular markers for constructing phylogenetic trees.
3. Phylogenetic Methods:
   - **Distance-Based Methods:** These methods measure genetic distances between sequences and construct trees based on the similarities observed. Common distance metrics include Jaccard, Hamming, and Jukes-Cantor distances.
   - **Maximum Likelihood:** This method estimates the evolutionary model that best fits the data and calculates the likelihood of observing the given sequences under that model. Maximum likelihood methods are widely used for their accuracy.
   - **Bayesian Inference:** Bayesian methods use probability distributions to estimate the evolutionary parameters. Bayesian phylogenetics incorporates prior knowledge and updates probabilities based on observed data, providing a robust framework for analysis.
4. Applications of Phylogenetics:
   - **Species Classification:** Phylogenetic analyses aid in classifying species and identifying evolutionary relationships, helping to refine taxonomy and clarify species boundaries.
   - **Biogeography:** By tracing evolutionary lineages, phylogenetics informs our understanding of the historical movement and distribution of species across continents and regions.
   - **Evolutionary Adaptations:** Studying the genetic changes along branches of a phylogenetic tree reveals adaptations and evolutionary innovations that have occurred in different lineages.
5. Challenges and Advances:
   - **Incomplete Lineage Sorting:** In some cases, gene trees within

a species do not match the species tree due to incomplete lineage sorting, posing challenges in reconstructing accurate evolutionary relationships.

- **Big Data and Computational Advances:** The increasing availability of genomic data demands sophisticated computational algorithms for efficient phylogenetic analyses. High-performance computing and machine learning techniques are being applied to handle large datasets.
- **Divergence Time Estimation:** Integrating fossil evidence and molecular data, scientists estimate the timing of evolutionary events, providing insights into the deep evolutionary history of organisms.

6. Future Directions:

- **Phylogenomics:** Integrating genomic data from multiple genes and organisms, phylogenomics provides a holistic view of evolutionary relationships, enhancing the accuracy of phylogenetic reconstructions.
- **Comparative Phylogenetics:** Comparative analyses across phylogenetic trees enable the study of evolutionary patterns, such as rates of speciation, adaptation, and extinction.
- **Environmental DNA (eDNA) and Metagenomics:** Environmental DNA analysis allows the study of entire ecosystems by examining genetic material present in environmental samples, advancing our understanding of microbial and ecological diversity.

In fact, phylogenetics plays a central role in elucidating the evolutionary history of life. Its applications span diverse fields, from understanding the relationships between species to uncovering the mechanisms driving evolutionary change. Continued advancements in molecular techniques and computational tools are enhancing our ability to unravel the complex and interconnected branches of the tree of life, providing profound insights into the biodiversity and evolutionary processes that have shaped our planet.

10.11.1. Tree

In the context of evolution, a "tree" often refers to a phylogenetic tree. A phylogenetic tree is a branching diagram or a tree showing the evolutionary relationships among various biological species or other entities based on their genetic, morphological, or behavioral similarities and differences. It depicts the evolutionary history of different species, illustrating how they have evolved from common ancestors over time.

Here are the key components and concepts related to phylogenetic trees in evolution theory:

1. Nodes and Branches:
   - **Nodes:** These represent common ancestors shared by the species or groups depicted in the tree.
   - **Branches:** These represent evolutionary relationships between the nodes. The length of branches can indicate the amount of evolutionary change.
2. Taxa:
   - **Terminal Nodes (Leaves):** Terminal nodes, often at the ends of the branches, represent living or extinct species or other entities being studied.
   - **Taxa:** Taxa (singular: taxon) refer to any group of species or organisms that are classified together, such as a genus, family, order, etc. Taxa are represented at the terminal nodes of the phylogenetic tree.
3. Rooted and Unrooted Trees:
   - **Rooted Trees:** These have a specific node designated as the root. The root represents the most recent common ancestor of all the taxa depicted in the tree.
   - **Unrooted Trees:** These do not have a specified root. They show the relationships between taxa without indicating a specific common ancestor.
4. Divergence Points:
   - **Divergence:** These points represent the points in evolutionary history where different species or groups split from a common ancestor. The more recent the divergence, the closer the relationship between the species.

5. Polytomy: It occurs when there is uncertainty about the exact branching order at a node. It indicates that the evolutionary relationships among the taxa are not fully resolved.

6. Evolutionary Distance:
   - **Branch Lengths:** In many phylogenetic trees, the lengths of the

branches represent the amount of evolutionary change (genetic or morphological) between the taxa. Longer branches typically indicate more evolutionary change.

7. Clades: These are groups of organisms that include an ancestor and all its descendants. Clades are depicted by nodes and their associated branches on a phylogenetic tree.

8. Homology and Convergence:

- **Homology:** Homologous traits are traits shared by different species due to their common ancestry. Homologous traits are crucial for constructing accurate phylogenetic trees.
- **Convergence:** Convergent evolution occurs when different species independently evolve similar traits due to similar environmental pressures. Convergent traits can complicate the interpretation of phylogenetic relationships.

Phylogenetic trees are fundamental tools in evolutionary biology, helping scientists understand the evolutionary history of life on Earth, the relationships between different species, and the processes of speciation and adaptation that have shaped biodiversity.

10.12. POLYMORPHISM

Polymorphism, in the context of evolutionary biology, refers to the coexistence of multiple forms (or phenotypes) within a population of a species. These different forms can be the result of genetic variation, allowing individuals within a species to exhibit diverse traits. Polymorphism is a key concept in evolutionary theory and has several important implications and forms:

1. Genetic Basis of Polymorphism:

- **Allelic Variation:** Polymorphism often arises from variations in specific genes (alleles) within a population. Different alleles can lead to variations in traits such as color, size, behavior, or resistance to diseases.
- **Gene Loci:** Polymorphism can occur at specific gene loci, where alternative forms of a gene (alleles) are present in the population.

2. Types of Polymorphism:

- **Single Nucleotide Polymorphism (SNP):** These are variations at a single DNA base pair. They are common and often used in genetic studies to understand the genetic basis of traits and diseases.

- **Insertion/Deletion Polymorphism (Indel):** Indels involve the insertion or deletion of a small DNA sequence. They can influence gene function and regulation.
- **Copy Number Variation (CNV):** These involve changes in the number of copies of specific DNA segments, leading to differences in gene dosage and expression levels.

3. Maintaining Polymorphism:
    - **Balancing Selection:** Polymorphism can be maintained in a population through various forms of balancing selection. Heterozygote advantage (or overdominance), where individuals with two different alleles have a higher fitness, can preserve polymorphism.
    - **Frequency-Dependent Selection:** The fitness of a particular phenotype depends on its frequency in the population. Rare phenotypes might have higher fitness, leading to their maintenance in the population.
4. Functional Significance:
    - **Adaptation:** Polymorphism can reflect adaptations to different environments. For example, different coat colors in animals might provide camouflage advantages in different habitats.
    - **Biotic Interactions:** Polymorphic traits can influence interactions with other organisms, such as predators, prey, or pathogens, leading to frequency-dependent selection.
5. Examples of Polymorphism:
    - **Mimicry:** Polymorphism is observed in species where some individuals mimic the appearance of other organisms (mimicry) to gain protection from predators or enhance hunting success.
    - **Sexual Dimorphism:** Differences in size, color, or other traits between males and females within a species represent a form of polymorphism related to sexual selection.
6. Human Evolution and Disease:
    - **Human Genetic Polymorphism:** Humans exhibit various genetic polymorphisms related to traits, diseases, and responses to medications.

- **Disease Resistance:** Certain genetic polymorphisms in human populations provide resistance to specific diseases such as malaria or HIV, demonstrating the evolutionary significance of polymorphism in the context of human health.

In summary, polymorphism is a fundamental concept in evolutionary biology, reflecting the genetic diversity within populations. It plays a crucial role in adaptation, ecological interactions, and the evolution of species over time. Studying polymorphism provides valuable insights into the mechanisms of evolution, natural selection, and the genetic basis of biodiversity.

10.13. PROTOCELL

Protocells are hypothetical precursors to modern cells that are believed to have played a crucial role in the origin of life on Earth. While protocells have not been observed directly, their study is fundamental in understanding the early stages of life's evolution. Here's a detailed and comprehensive overview of protocells in evolutionary biology:

1. Definition and Characteristics:

- **Definition:** A protocell is a self-organized, cell-like structure that exhibits some properties of living cells but lacks the full complexity of modern cells.
- **Basic Characteristics:** Protocells are thought to possess a boundary (often lipid-based), internal compartments, and the potential to maintain a chemical gradient across their boundary.

2. Origin of Life Hypotheses:

- **Primordial Soup Theory:** Protocells might have formed in a nutrient-rich "soup" of organic molecules, where simple molecules gradually organized into more complex structures.
- **Hydrothermal Vent Theory:** Some theories propose that protocells could have originated near hydrothermal vents on the ocean floor, where mineral-laden water provides a unique environment for chemical reactions.

3. Components and Functions:

- **Lipid Bilayers:** Protocell membranes are often lipid-based, similar to modern cell membranes. These lipid bilayers can form spontaneously in water and enclose other molecules.
- **RNA World Hypothesis:** RNA, a molecule capable of both

information storage and catalysis, is often associated with protocells. It might have played a role in early genetic information and enzymatic functions.

- **Metabolic Reactions:** Protocells are theorized to have undergone simple metabolic reactions, allowing them to generate energy and carry out basic functions.

4. Self-Organization and Complexity:
   - **Emergence of Complexity:** Protocells represent a critical step in the evolution of complexity, where simple chemical systems started exhibiting lifelike properties.
   - **Self-Organization:** Protocells are examples of self-organization, where simple components organize themselves into structures with functional properties, a phenomenon essential in the evolution of life.
5. Experimental Approaches:
   - **Laboratory Experiments:** Scientists conduct experiments to recreate protocell-like structures in the laboratory. These experiments involve lipid vesicles, chemical gradients, and catalytic molecules to simulate protocell behaviors.
   - **Artificial Life Studies:** Computer simulations and artificial life experiments help researchers model and understand the conditions under which protocells could form and evolve.
6. Protocells and Evolutionary Transitions:
   - **Transition to Cellular Life:** Protocells are considered a bridge between simple molecules and fully functional cellular life forms. Studying protocells informs our understanding of the transition from non-living to living entities.
   - **Evolutionary Implications:** Protocells provide insights into the evolutionary processes that led to the development of cellularity, replication, metabolism, and the emergence of genetic information systems.
7. Challenges and Future Directions:
   - **Origin of Complexity:** Understanding how protocells transitioned from simple chemical systems to entities capable of replication, metabolism, and response to the environment is a major challenge.

- **Laboratory Challenges:** Recreating realistic protocell conditions in the lab, including the right chemical environment and energy sources, presents significant challenges.

8. Astrobiology and Extraterrestrial Life:

- **Search for Protocells:** Astrobiologists explore the possibility of protocells or similar structures existing on other planets or moons in our solar system, especially where conditions might support simple organic chemistry.

So, protocells represent a critical stage in the evolutionary process, where self-organization and the emergence of lifelike properties began to pave the way for the complexity of modern living organisms. Studying protocells offers profound insights into the origins of life on Earth and provides a framework for understanding the potential for life elsewhere in the universe. Ongoing research in this field continues to illuminate the pathways that led to the emergence of life as we know it.

10.14. SYSTEMATICS

Systematics is the scientific study of the diversity of organisms, both past and present, and the relationships among them. It encompasses the fields of taxonomy, phylogenetics, and evolutionary biology, aiming to classify organisms based on evolutionary relationships and understand the evolutionary history of life on Earth. Here's a detailed overview of systematics:

1. Taxonomy:

- **Definition:** Taxonomy is the science of naming, defining, and classifying organisms into hierarchical categories, including species, genera, families, orders, classes, phyla, and kingdoms.
- **Taxonomic Ranks:** Each rank represents a level of classification, with species being the fundamental unit. Taxonomists use morphological, genetic, and ecological data to classify organisms.

2. Phylogenetics:

- **Definition:** Phylogenetics is the study of evolutionary relationships among organisms. It involves constructing phylogenetic trees or networks to represent the evolutionary history and ancestry of different species.
- **Molecular Phylogenetics:** Molecular data, such as DNA and

protein sequences, are widely used in phylogenetic analyses to infer evolutionary relationships and understand genetic divergence.

3. Cladistics:
    - **Definition:** Cladistics is a method of classification that groups organisms based on shared derived characteristics, or synapomorphies. Cladistic analyses result in cladograms, branching diagrams showing evolutionary relationships.
4. Evolutionary Taxonomy:
    - **Definition:** Evolutionary taxonomy integrates both evolutionary relationships and traditional morphological characteristics into classification. It emphasizes the importance of common ancestry in taxonomy.
5. Biodiversity Informatics:
    - **Definition:** Biodiversity informatics involves the application of computational methods to manage, analyze, and visualize biological data. It includes databases, geographic information systems (GIS), and computational tools to study biodiversity patterns.
6. Biogeography:
    - **Definition:** Biogeography is the study of the distribution of organisms across geographic space and through geological time. It explores the factors influencing species distribution and the historical movements of species.
7. Applications of Systematics:
    - **Conservation Biology:** Systematics helps identify and prioritize endangered species for conservation efforts by understanding their evolutionary uniqueness and relationships.
    - **Biomedical Research:** Understanding the evolutionary relationships among organisms' aids in biomedical research, including the study of disease vectors and evolutionary medicine.
    - **Evolutionary Ecology:** Systematics contributes to our understanding of ecological interactions and evolutionary adaptations in specific lineages of organisms.
8. Challenges and Future Directions:
    - **Rapid Taxonomic Changes:** Advances in molecular techniques

often lead to rapid changes in the classification of organisms. Keeping taxonomies updated poses a challenge.

- **Big Data and Computational Advances:** Managing large biological datasets and applying computational methods are ongoing challenges and opportunities in systematics research.

9. Ethical and Societal Implications:

- **Biodiversity Conservation:** Systematics plays a vital role in biodiversity conservation and informs policy decisions related to habitat protection and species conservation efforts.
- **Ethical Considerations:** Systematics research raises ethical questions about the rights of indigenous communities, ethical collection of specimens, and the implications of genetic research.

In summary, systematics is a dynamic field that continues to evolve with technological advancements and a deeper understanding of evolutionary processes. It provides the foundation for our knowledge of life's diversity and evolutionary history, offering valuable insights into the interconnectedness of all living organisms on Earth.

10.15. TRANSGENERATIONAL EPIGENETIC INHERITANCE

Transgenerational epigenetic inheritance refers to the transmission of epigenetic information from one generation to the next, influencing traits and characteristics in offspring without changes to the underlying DNA sequence. Unlike genetic inheritance, which involves alterations to the DNA sequence, epigenetic inheritance involves changes in gene activity or expression patterns that can be influenced by environmental factors. Here's a detailed overview of transgenerational epigenetic inheritance:

1. Epigenetic Mechanisms:

- **DNA Methylation:** One of the most studied epigenetic modifications involves the addition of methyl groups to DNA, which can silence gene expression.
- **Histone Modifications:** Chemical modifications to histone proteins can alter the structure of chromatin, affecting how genes are expressed.
- **Non-Coding RNAs:** Small RNAs, such as microRNAs, can regulate gene expression by targeting specific messenger RNAs

for degradation or by inhibiting their translation into proteins.

2. Environmental Influence:
   - **Early Life Experiences:** Environmental factors during early development, such as maternal diet, stress, or exposure to toxins, can lead to epigenetic changes that affect the health and development of offspring.
   - **Multigenerational Effects:** Epigenetic changes induced by environmental exposures can sometimes be passed on to subsequent generations, potentially affecting the phenotype and disease susceptibility of descendants.
3. Transgenerational Epigenetic Inheritance in Humans and Animals:
   - **Human Studies:** Research suggests that certain environmental exposures, such as famine or stress during pregnancy, may lead to epigenetic changes that affect multiple generations.
   - **Animal Studies:** In animals, transgenerational epigenetic inheritance has been observed in various species, including mice and rats, demonstrating the heritability of acquired traits.
4. Mechanisms of Transmission:
   - **Gametic Inheritance:** Epigenetic information can be transmitted through the germline, affecting the development of eggs and sperm, and subsequently influencing the traits of offspring.
   - **Somatic Inheritance:** Epigenetic changes can also be inherited through somatic cells, potentially affecting specific tissues or organs in subsequent generations.
5. Evolutionary Implications:
   - **Rapid Adaptation:** Transgenerational epigenetic inheritance may provide a mechanism for rapid adaptation to changing environments, allowing populations to respond to environmental challenges without changes to the DNA sequence.
   - **Evolutionary Conservation:** Certain epigenetic mechanisms and patterns are conserved across species, suggesting their evolutionary significance in shaping phenotypic diversity and adaptation.
6. Challenges and Future Research:
   - **Epigenetic Stability:** The stability of transgenerational epigenetic

changes and their persistence across multiple generations are active areas of research and debate.

- **Molecular Understanding:** Understanding the molecular mechanisms underlying transgenerational epigenetic inheritance, including how specific epigenetic marks are established and maintained, is a focus of ongoing research.

In conclusion, transgenerational epigenetic inheritance highlights the complexity of heredity beyond the DNA sequence, emphasizing the role of epigenetic modifications in shaping phenotypic variation and adaptation across generations. This emerging field continues to unravel the mechanisms and implications of epigenetic inheritance, providing insights into evolution, development, and disease susceptibility in diverse organisms, including humans.

10.16. OVERVIEW OF EVOLUTIONARY APPLICATIONS

Evolutionary biology, beyond its foundational role in understanding life's history, has practical applications in various fields. Here's an overview of the diverse applications of evolutionary principles:

1. Medicine and Public Health:

- **Understanding Disease:** Evolutionary biology helps understand the evolution of diseases, including antibiotic resistance in bacteria and the evolution of viruses. This knowledge aids in the development of effective treatments.
- **Personalized Medicine:** Genetic variation among individuals, a product of evolution, influences responses to treatments. Understanding these variations enables personalized medicine approaches.

2. Agriculture and Crop Improvement:

- **Crop Domestication:** Evolutionary insights aid in understanding the domestication processes of crops, leading to improved agricultural practices and the development of new cultivars.
- **Resistance Breeding:** Studying the evolution of plant pests and diseases guides the development of resistant crop varieties, ensuring food security.

3. Conservation Biology:

- **Conservation Genetics:** Evolutionary genetics helps assess genetic diversity within endangered populations. Conservation strategies use this information to preserve genetic variability and prevent inbreeding.
- **Assisted Evolution:** Evolutionary principles inform assisted evolution strategies, including selective breeding and translocation, to enhance species' resilience to changing environments.

4. Environmental Management and Restoration:

- **Ecosystem Restoration:** Evolutionary considerations are essential in restoring ecosystems. Restoring native species adapted to specific environments ensures ecological resilience and biodiversity.
- **Invasive Species Management:** Understanding evolutionary processes in invasive species aids in developing effective management strategies, preventing ecological disruption.

5. Forensic Biology:

- **Forensic Genetics:** Evolutionary patterns in human DNA inform forensic investigations. Genetic markers help identify individuals and trace evolutionary relationships among populations.

6. Biotechnology and Industrial Applications:

- **Bioprospecting:** Studying evolutionary adaptations in organisms leads to the discovery of novel enzymes and molecules with applications in biotechnology, pharmaceuticals, and industrial processes.
- **Bioinformatics:** Evolutionary algorithms inspire computational techniques used in bioinformatics for sequence analysis, structural prediction, and drug design.

7. Understanding Human Evolution:

- **Human Health:** Knowledge of human evolutionary history aids in understanding genetic predispositions to diseases, immune system evolution, and human adaptations to different environments.
- **Social Sciences:** Evolutionary psychology and anthropology provide insights into human behavior, cooperation, and cultural evolution.

8. Education and Public Understanding:

- **Science Communication:** Teaching evolutionary principles enhances scientific literacy and promotes understanding of biological concepts among the public.
- **Countering Misconceptions:** Evolutionary education helps counter misconceptions, ensuring accurate scientific knowledge in society.

So, evolutionary applications have a far-reaching impact, from guiding medical treatments to conserving biodiversity and managing ecosystems. Understanding evolution not only enriches our knowledge of the natural world but also informs practical solutions to real-world challenges across diverse disciplines.

REFERENCES

1. Bowler, P. J., (1989). *Evolution: The History of an Idea*. University of California Press, Berkeley.
2. Darwin, C. R., (1859). *The Origin of Species by Means of Natural Selection, or the Preservation of Favored Races in the Struggle for Life* (1st edn.). London: John Murray. [Darwin 1859 available online].
3. Dobzhansky, T., (1973). Nothing in biology makes sense except in the light of evolution. *American Biology Teacher, 35*(3), 125–129. doi: 10.2307/4444260.
4. Endler, J., (1986). *Natural Selection in the Wild*. Princeton, NJ: Princeton University Press.
5. Futuyma, D. J., (1986). *Evolutionary Biology*. Sinauer Associates, inc. Sunderland, Mass.
6. Futuyma, D. J., (2005). *Evolution*. Sunderland, MA: Sinauer Associates.
7. Glass, B., Temkin, O., & Strauss, W. L., (1959). *Forerunners of Darwin: 1745–1859*. Johns Hopkins Press, Baltimore.
8. Milner, R., (1990). *The Encyclopedia of Evolution: An Owl Book*. Henry Holt and Company. New York.
9. *The Descent of Man, and Selection in Relation to Sex* (2nd edn.). London: John Murray. [Darwin 1874 available online].
10. Van Valen, L., (1973). Book review: Festschrift for George Gaylord Simpson. *Science, 180*(4085), 488. doi: 10.1126/science.180.4085.486.

CHAPTER 11

Controversies and Debates

CONTENTS

11.1. CREATIONISM VS. EVOLUTION

Creationism and evolution represent two distinct viewpoints regarding the origins of life and the development of species on Earth. Here's a comparative analysis of these contrasting perspectives (Figure 11.1):

Figure 11.1. Controversies and debates in evolution theory.

Source: By Mike Licht – Flickr: Anti-Evolution League, at the Scopes Trial, Dayton Tennessee, CC BY 2.0, https://commons.wikimedia.org/w/index.php?curid=32160684; Creative Commons License.

1. Creationism:

i. Definition: Creationism is the belief that life and the universe were created by a divine being or deity, often based on religious texts. Creationists typically adhere to a literal interpretation of religious scriptures, such as the Bible's Genesis account.

ii. Central Tenets:

- **Creation Ex Nihilo:** Creationists believe that life was created out of nothing by a supernatural force, as described in religious texts.
- **Young Earth Creationism:** Some creationists assert that the Earth and all life forms were created in a literal six-day period around 6,000 to 10,000 years ago, based on genealogical timelines mentioned in religious texts.
- **Critiques of Evolution:** Creationists often challenge the scientific validity of evolutionary theory, questioning aspects like the fossil record, irreducible complexity, and the origin of life.

2. Evolution:

i. Definition: Evolution is the scientific theory that explains the diversity of life on Earth through processes such as natural selection, genetic variation, and speciation. It is supported by extensive empirical evidence from multiple scientific disciplines, including paleontology, genetics, and comparative anatomy.

ii. Central Tenets:

- **Common Ancestry:** Evolution proposes that all living organisms share a common ancestry, with species diverging and adapting over millions of years.
- **Natural Selection:** Evolutionary changes occur through natural selection, where traits that enhance an organism's fitness in a given environment are passed on to subsequent generations.
- **Scientific Consensus:** Evolution is widely accepted within the scientific community and forms the foundation of modern biology. It is supported by a vast body of evidence, including fossil records, genetic similarities among species, and observed instances of microevolution and speciation.

3. Points of Conflict:

i. **Interpretation of Religious Texts:** Creationism relies on religious texts for its assertions, often conflicting with scientific findings when there is a literal interpretation of these texts.

ii. **Methodology and Empirical Evidence:** Evolution is based on empirical evidence, observation, and experimentation. Creationism, being rooted in faith, does not adhere to the scientific method in the same way and lacks empirical evidence.

4. Attempts at Synthesis:

i. **Theistic Evolution:** Some individuals and religious groups reconcile faith with evolutionary theory by adopting theistic evolution. This perspective asserts that a higher power guided the process of evolution, aligning religious belief with scientific understanding.

5. Education and Societal Impact:

i. **Education:** The teaching of creationism in public schools has been a topic of controversy, with legal battles in various countries, such as the United States, regarding its inclusion in science curricula.

ii. **Cultural and Political Influence:** Differences in belief regarding creationism and evolution can influence public opinion, political debates, and policies related to education and science.

In true sense, creationism and evolution represent fundamentally different approaches to understanding the origins of life. While evolution is grounded in scientific evidence and forms the basis of modern biology, creationism is rooted in religious belief. The coexistence of these viewpoints continues to be a topic of debate, raising questions about the relationship between faith, science, and education in society.

11.2. INTELLIGENT DESIGN

Intelligent Design (ID) and Evolution Theory are two contrasting perspectives that address the origins and development of life on Earth. Here's a detailed overview of both concepts:

1. Intelligent Design (ID):

i. Definition: Intelligent Design is the concept that certain features of the natural world are best explained by an intelligent cause rather than an undirected process such as natural selection. It proposes that life's complexity and specific biological structures are evidence of purposeful design.

ii. Central Tenets:

- **Irreducible Complexity:** ID proponents argue that certain biological systems are too complex to have evolved through gradual modifications and must have been designed in their complete form.
- **Specified Complexity:** ID advocates claim that some biological structures exhibit patterns that are highly improbable and exhibit specified complexity, suggesting the involvement of an intelligent designer.
- **Critiques of Evolution:** ID proponents often criticize aspects of evolutionary theory, including the origin of life, the complexity of biological systems, and the lack of transitional forms in the fossil record.

2. Evolution Theory:

i. Definition: Evolution is the scientific theory that explains the diversity of life through processes like natural selection, genetic variation, mutation, and speciation. It posits that all living organisms share a common ancestry and have evolved over millions of years.

ii. Central Tenets:

- **Common Descent:** Evolution proposes that all living organisms are descended from a common ancestor, with species diversifying over time.
- **Natural Selection:** Evolutionary changes occur through natural selection, where organisms with advantageous traits are more likely to survive and reproduce, passing those traits to the next generation.
- **Scientific Support:** Evolution is supported by extensive empirical evidence, including the fossil record, comparative anatomy, molecular biology, and observed instances of natural selection and speciation.

3. Points of Conflict:

i. **Complexity and Design:** The main conflict between ID and evolution lies in the interpretation of complexity. ID argues that complexity implies design, while evolution explains complexity through natural processes like natural selection.

ii. **Scientific Acceptance:** Evolution is widely accepted within the scientific community and is considered a foundational theory in biology. ID, however, is not widely accepted as a scientific theory due to its lack of empirical evidence and testable hypotheses.

4. Attempts at Synthesis:

i. **Theistic Evolution:** Some individuals reconcile faith with evolutionary theory by embracing theistic evolution. This perspective asserts that a higher power, often seen as the divine, guided the process of evolution, allowing for religious beliefs and scientific understanding to coexist.

5. Educational Controversy:

i. Public Education: The teaching of ID in public schools has been a subject of controversy, especially concerning its inclusion in science curricula. Legal battles have taken place in various countries over the inclusion of ID alongside evolution in science classrooms.

6. Cultural and Societal Impact:

i. **Public Opinion:** Differences in beliefs about ID and evolution can influence public opinion, leading to debates about science education, public policy, and the relationship between science and faith in society.

Intelligent Design and Evolution Theory represent contrasting viewpoints on the origins and complexity of life. While evolution is rooted in extensive scientific research and forms the foundation of modern biology, Intelligent Design relies on the notion of purposeful design. These differing perspectives continue to spark debates and discussions in the realms of science, education, and public discourse.

11.3. DEBATES WITHIN THE SCIENTIFIC COMMUNITY

Within the scientific community, evolution theory is widely accepted as the foundational framework for understanding the diversity of life on Earth. However, this acceptance doesn't mean that the theory is without debate or controversy. Here's a detailed overview of the debates within the scientific community about controversies and debates regarding evolution theory:

1. Debates about Mechanisms:

i. **Punctuated Equilibrium vs. Gradualism:** There is a debate about the pace of evolution. Punctuated equilibrium suggests that evolutionary change occurs rapidly in relatively short periods, while gradualism posits that evolution happens steadily over long periods.

2. Debates about Evolutionary Patterns:

i. **Mass Extinctions:** The debate revolves around the impact of mass extinctions on the course of evolution. Some scientists argue that these events play a significant role in shaping biodiversity, while others emphasize the importance of gradual, continuous change.

3. Debates about Molecular Evolution:

i. **Neutral Theory vs. Selection:** There is a debate about the significance of neutral mutations in molecular evolution. Neutral theory suggests that many genetic changes are neutral and do not affect fitness, while the selectionist perspective emphasizes the role of natural selection in shaping molecular evolution.

4. Debates about Evolutionary Developmental Biology (Evo-Devo):

i. **Role of Developmental Genes:** Scientists debate how changes in developmental genes influence evolutionary processes. Evo-Devo explores how alterations in genes controlling development lead to morphological diversity.

5. Debates about the Evolution of Complexity:

i. **Origin of Complexity:** There are debates about how complex traits and structures, such as the vertebrate eye, originated. Some scientists argue for stepwise evolution, while others propose irreducible complexity, suggesting certain structures cannot evolve through gradual steps.

6. Debates about Human Evolution:

i. **Interpreting Fossil Evidence:** Scientists debate the interpretation of fossil evidence related to human evolution, including the identification of ancestors and the timing of key evolutionary events.

ii. **Genetic Ancestry:** Advances in genetics have led to debates about the genetic ancestry of modern humans, with ongoing research into the relationships between different hominin species.

7. Debates about Evolutionary Psychology:

i. **Human Behavior and Evolution:** Evolutionary psychology explores the influence of evolutionary processes on human behavior. Debates arise concerning the extent to which specific behaviors are adaptations shaped by evolution.

8. Debates about Evolutionary Medicine:

i. **Mismatch Hypothesis:** Evolutionary medicine explores the impact of past evolutionary environments on modern human health. Debates focus on the extent to which diseases result from a mismatch between current lifestyles and ancestral environments.

9. Debates about Teaching Evolution:

i. **Science Education:** There are debates about how evolution should be taught in schools, with ongoing discussions regarding the inclusion of creationism or Intelligent Design in science curricula.

In short, debates within the scientific community about evolution theory are intrinsic to the scientific process. These debates fuel ongoing research, leading to a deeper understanding of evolutionary mechanisms, patterns, and their implications for diverse fields of study. Scientific disagreements and debates, when conducted rigorously and respectfully, contribute to the advancement of knowledge in the field of evolutionary biology.

11.4. TEACHING EVOLUTION IN SCHOOLS

Teaching evolution in schools is a topic of considerable debate and importance, involving educational, scientific, and sometimes, political, and religious considerations. Here's a comprehensive overview of the various aspects related to teaching evolution in schools:

1. Scientific Consensus:

i. **Foundation of Biology:** Evolution is a foundational concept in biology and is widely accepted within the scientific community. It provides the framework for understanding the diversity of life, adaptation, and the interconnectedness of all living organisms.

ii. **Empirical Evidence:** Evolutionary theory is supported by extensive empirical evidence from diverse fields such as paleontology, genetics, comparative anatomy, and molecular biology.

2. Challenges and Controversies:

i. **Religious and Cultural Beliefs:** Some religious and cultural beliefs conflict with the teaching of evolution, leading to challenges in regions where these beliefs are prevalent.

ii. **Public Controversy:** The teaching of evolution has been a contentious issue in public education, leading to legal battles and debates about the inclusion of alternative views, such as creationism or Intelligent Design, in science classrooms.

3. Legal Framework:

i. **Supreme Court Decisions:** In the United States, landmark Supreme Court decisions, such as the 1987 Edwards v. Aguillard case, have ruled that teaching creationism in public schools is unconstitutional due to the separation of church and state.

ii. **Scientific Standards:** Many countries have established scientific standards for education that include the teaching of evolution as a fundamental concept in science education.

4. Pedagogical Approaches:

i. **Age-Appropriate Teaching:** Educators often tailor the teaching of evolution based on the grade level, ensuring that the complexity of the topic aligns with students' cognitive abilities.

ii. **Hands-on Learning:** Engaging students in activities like examining fossils or understanding natural selection through simulations enhances their comprehension and acceptance of evolutionary concepts.

5. Teaching Controversies:

i. **Teaching Controversial Aspects:** Educators sometimes avoid or struggle with teaching controversial aspects of evolution, such as human evolution, due to concerns about parental or community objections.

ii. **Teacher Preparation:** Teachers need adequate training and support to navigate potential conflicts in the classroom and present evolutionary concepts effectively.

6. International Perspectives:

i. **Global Variations:** The acceptance and teaching of evolution vary globally. Some countries seamlessly integrate evolutionary theory into their curricula, while others face significant resistance due to religious or cultural factors.

7. Educational Outreach and Public Understanding:

i. **Community Engagement:** Efforts to educate parents and communities about the scientific importance of evolution can foster a more supportive environment for the teaching of evolution in schools.

ii. **Scientific Literacy:** Promoting scientific literacy is essential to ensuring that the public understands the foundational role of evolution in modern biology.

8. Future Challenges and Prospects:

i. **Advancements in Education:** Integrating modern teaching methods, such as interactive digital resources and citizen science projects, can enhance students' understanding and acceptance of evolutionary concepts.

ii. **Addressing Misconceptions:** Teachers and educators continue to work on addressing common misconceptions about evolution, promoting critical thinking and evidence-based reasoning.

In summary, teaching evolution in schools requires careful navigation of cultural, religious, and educational landscapes. Effective education in evolutionary biology not only imparts scientific knowledge but also fosters critical thinking skills and an understanding of the scientific process, crucial for informed citizenship in our increasingly complex world.

3

REFERENCES

1. *ARISTOTLE, The Physics*, (1957). London: William Heinemann Ltd.
2. Barton, N. H., Derek, E. G. B., Jonathan, A. E., David, G., & Nipam, H. P., (2007). *Evolution*. Cold Spring Harbor, NY: Cold Spring Harbor Laboratory Press.
3. Boolootian, R. A., (1981). *College Zoology* (11th edn.). London: Collier Macmillan.
4. Costa, J. C., (2009). *The Annotated Origin of Species*. Cambridge, MA: Harvard Univ. Press.
5. Coyne, J. A., (2009). *Why Evolution is True*. Oxford: Oxford Univ. Press.
6. Criswell, W. A., (1980). *Did Man Just Happen?* Chicago: Moody Press.
7. Darwin, C., (1968). *On the Origin of Species*. London: Penguin Books.
8. *Divine Principle*, (1973). (2nd edn.). New York: Holy Spirit Association for the Unification of World Christianity.
9. Dobzhansky, T., (1960). *Evolution of Life* (p. 40). Chicago: University of Chicago Press. "Scientific American." In Duncan–1978.
10. Dobzhansky, T., (1973). Nothing in biology makes sense except in the light of evolution. *American Biology Teacher, 35*(3), 125–129.
11. Duncan, H., (1978). *Evolution: The Incredible Hoax*. Lubbock, Texas: Missionary Crusade.
12. Engels, F., (1970). *Speech at the Graveside of Karl Marx*. In Marx and Engels, Selected Works. Moscow: Progress Publishers.
13. Freeman, S., & Jon, C. H., (2014). *Evolutionary Analysis* (5th edn.). San Francisco: Benjamin Cummings.
14. Futuyma, D. J., & Mark, K., (2017). *Evolution* (4th edn.). Sunderland, MA: Sinauer Associates.
15. Futuyma, D. J., (1986). *Evolutionary Biology* (2nd edn.). Sunderland, Massachusetts: Sinauer Associates Inc.
16. Gish, D. T., (1972). *Speculations and Experiments Related to Theories on the Origin of Life (A Critique)*. San Diego: Creation-Life Publishers.
17. Gould, S. J., (1980). *Ever Since Darwin*. New York: W.W. Norton & Company–1977. The Panda's Thumb. New York: W.W. Norton & Company.
18. Heinze, T. F., (1973). *Creation vs. Evolution Handbook*. Grand Rapids, Michigan: Baker Book House.

19. Hitching, F., (1982). *The Neck of the Giraffe*. Ontario: The New American Library of Canada.
20. Hoyle, F., & Chandra, W. N., (1981). *Evolution from Space: A Theory of Cosmic Creationism*. New York: Simon and Schuster.
21. Huxley, J. S., (1963). *Evolution in Action*. Great Britain: Penguin Books.
22. Imanishi, K., (1976). *What is Evolution? (in Japanese) Tokyo: Kodansha*.
23. Komatsu, S., (1982). *A Nonstandardized Biology (in Japanese)*. Tokyo: Shinchosha.
24. Losos, J., (2014). *The Princeton Guide to Evolution*. Princeton, NJ: Princeton Univ. Press.
25. Mayr, E., (1963). *Animal Species and Evolution*. Cambridge, MA: Harvard University Press.
26. Miller, J. D., Scott, E. C., & Okamoto, S., (2006). Public acceptance of evolution. *Science, 313*, 765–766.
27. Monod, J., (1971). *Chance and Necessity*. New York: Random House Inc.
28. Moon, S. M., (1991). *The Unification of the World and Head-Wing Thought, or Godism*. (Sermon given at the Unification Church Headquarters of Seoul on August 25, 1991). Family Magazine (in Japanese).
29. Moore, R., (1969). *Evolution (Japanese Edition)*. Tokyo: Time-Life International, (originally published by Time Inc., New York, 1964).
30. Myednykov, B. M., (1982). *The Frontier of the Interesting Theory of Evolution (Japanese Edition)*. Tokyo: Tokyo Tosho Co., Ltd.
31. Nagano, K., (1986). *The Frontline of Biology (in Japanese)*. Tokyo: Nihon Hyoronsha.
32. Nakahara, H., & Takashi, S., (1986). *Why Does Man Evolve? (in Japanese)*. Tokyo Tairyusha.
33. Nelson, B., (1967). *After Its Kind*. Minneapolis: Bethany House Publisher.
34. Ruse, M., & Joseph, T., (2009). *Evolution: The First 4 Billion Years*. Cambridge, MA: Harvard Univ. Press.
35. Suzuki, H., & Hideo, M., (1987). *Explaining New Biology (in Japanese)*. Tokyo: Bun-Eido.

36. Takei, K., (1987). *The Myth of the Archaeopteryx on the Verge of Falling (in Japanese)*. Kagaku-Asahi Magazine. Tokyo: Asahi-Shimbun Sha.
37. Ueda, R., & Hiroshi, S., (1974). *Explaining Biology II (in Japanese)*. Tokyo: Bun-Eido.
38. Usubuchi, I., (1985). *The Theory of Evolution Reexamined (in Japanese)*. Tokyo: Kodansha.
39. Watson, L., (1979). *Lifetide*. London: Hodder and Stoughton.
40. Whitcomb, J. C. Jr., (1972). *The Early Earth*. Grand Rapids, Michigan: Baker Book House.
41. Zimmer, C., & Douglas, J. E., (2016). *Evolution: Making Sense of Life* (2nd edn.). Colorado: Roberts.

CHAPTER 12

Future Directions in Evolutionary Biology

CONTENTS

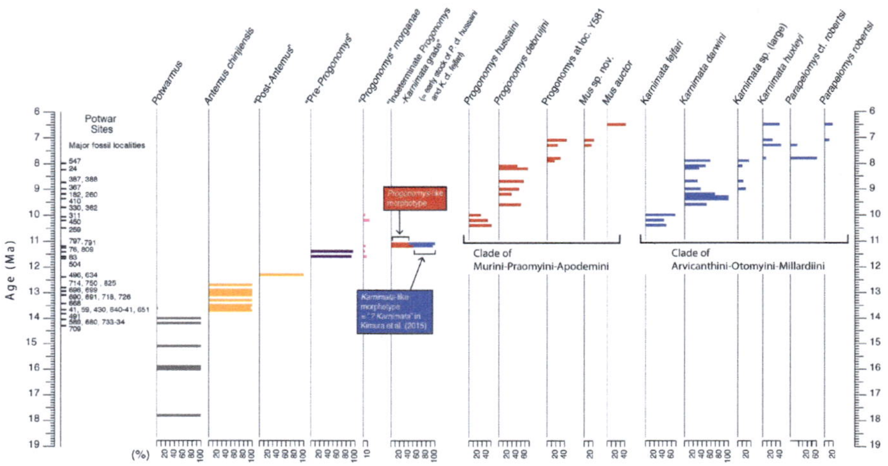

Figure 12.1. Future directions in evolutionary biology.

Source: Kimura, Y., Flynn, L. J., & Jacobs, L. L., (2021). Tempo and mode: Evidence on a protracted split from a dense fossil record. Front. Ecol. Evol., 9, 642814. doi: 10.3389/fevo.2021.642814; Creative Commons License.

12.1. ADVANCES IN GENOMICS AND BIG DATA

Advances in genomics and the handling of big data have revolutionized the field of evolutionary biology, providing unprecedented insights into the genetic basis of evolution, species relationships, and ecological interactions. Here's a detailed overview of how genomics and big data are shaping the future of evolutionary biology:

1. Genomic Sequencing:

   i. **High-Throughput Sequencing:** Next-generation sequencing technologies enable the rapid and cost-effective sequencing of entire genomes, facilitating large-scale comparative genomics across diverse species.

   ii. **Reference Genomes:** Efforts are underway to create comprehensive reference genomes for a wide array of organisms, providing a foundation for understanding genetic variation within and between species.

2. Population Genomics:

   i. **Genetic Variation Studies:** Large-scale sequencing of populations allows the identification of genetic variations, including single nucleotide polymorphisms (SNPs) and structural variants, shedding light on evolutionary processes like adaptation and genetic drift.

ii. **Phylogeography:** Genomic data enhance phylogeographic studies, revealing historical migration patterns, population dynamics, and the impact of geography on genetic diversity.

3. Comparative Genomics:

i. **Gene Family Evolution:** Comparative genomics elucidates the evolution of gene families, gene duplication events, and the functional divergence of genes, providing insights into the genetic basis of phenotypic diversity.

ii. **Conserved Elements:** Identification of conserved non-coding elements (CNEs) and regulatory regions aids in understanding the evolution of gene regulation and the role of cis-regulatory changes in adaptation.

4. Evolutionary Transcriptomics:

i. **RNA-Seq Studies:** Transcriptomic analyses provide information on gene expression patterns across different tissues and developmental stages, facilitating the understanding of evolutionary changes in gene regulation.

ii. **Alternative Splicing:** Studying alternative splicing events reveals how transcriptomic diversity contributes to phenotypic variation and adaptation in different environments.

5. Big Data Analytics:

i. **Machine Learning:** These algorithms process vast biological datasets, identifying patterns and correlations that are often challenging for human researchers to discern, leading to new discoveries in evolutionary processes.

ii. **Network Analyses:** Big data approaches enable the construction of intricate biological networks, uncovering gene interactions, metabolic pathways, and co-evolutionary relationships between species.

6. Ancient DNA Studies:

i. **Paleogenomics:** Advances in ancient DNA extraction and sequencing techniques allow the analysis of ancient genomes. Studying ancient DNA provides insights into extinct species, evolutionary relationships, and adaptation to past environmental changes.

7. Integrative Biology:
   - i. **Multi-Omics Approaches:** Integration of genomics with other 'omics' disciplines, such as proteomics, metabolomics, and epigenomics, offers a holistic view of evolutionary processes, considering genetic, biochemical, and environmental factors.
   - ii. **Eco-Evolutionary Dynamics:** Combining genetic data with ecological parameters enhances our understanding of eco-evolutionary dynamics, including species interactions, adaptation to changing environments, and the impact of evolution on ecosystems.

8. Ethical and Societal Implications:
   - i. **Data Privacy:** As biological data collection expands, addressing ethical concerns related to data privacy, consent, and responsible data usage becomes crucial to maintain public trust and ensure the ethical conduct of research.
   - ii. **Public Engagement:** Educating the public about the benefits and potential risks associated with genomic research fosters a scientifically literate society and encourages informed discussions about ethical considerations.

Thus, the integration of genomics and big data analytics is propelling evolutionary biology into an era of unprecedented discovery and understanding. These advances not only deepen our knowledge of evolutionary processes but also have profound implications for fields ranging from medicine and agriculture to conservation and ecology, shaping the future of biological research and its applications.

12.2. EVOLUTION IN THE ANTHROPOCENE

The Anthropocene epoch, characterized by the profound impact of human activities on Earth's ecosystems, has ushered in unique evolutionary challenges and opportunities for living organisms. Here's a detailed overview of how evolution operates in the Anthropocene:

1. Rapid Adaptation to Environmental Changes:
   - i. **Climate Change:** Species are evolving to cope with changing climate patterns, including altered temperature and precipitation

regimes. Rapid adaptation is observed in both plant and animal populations.

ii. **Urbanization:** Urban environments create novel selection pressures. Some species adapt to city life, exhibiting behaviors and traits that enhance survival in urban landscapes.

2. Human-Induced Selection Pressures:

i. **Pollution:** Evolutionary responses to pollutants, such as pesticides and heavy metals, are evident in various organisms. Some species develop resistance, while others face population declines.

ii. **Selective Harvesting:** Overfishing and hunting lead to the selective removal of individuals. This can induce rapid changes in the genetic makeup of targeted populations.

3. Invasive Species and Hybridization:

i. **Invasive Species:** Invasive species, introduced by human activities, can outcompete native species. Rapid evolution in both invasive and native species occurs as they adapt to new competitive environments.

ii. **Hybridization:** Human-induced habitat changes lead to increased interactions between species that wouldn't naturally meet. Hybridization events can lead to the emergence of new genetic lineages.

4. Evolutionary Consequences for Biodiversity:

i. **Extinction and Speciation:** The Anthropocene has witnessed numerous extinctions due to habitat destruction and climate change. Simultaneously, new species are emerging through adaptation to human-altered environments.

ii. **Biodiversity Loss:** Human-induced changes often lead to homogenization of ecosystems, resulting in the loss of unique genetic diversity within species.

5. Coevolution with Humans:

i. **Pathogen Evolution:** Pathogens, including bacteria and viruses, evolve in response to human interventions such as antibiotics. This has led to the emergence of antibiotic-resistant strains.

ii. **Agricultural Evolution:** Domesticated plants and animals are under strong selective pressure due to human agriculture practices, resulting in changes in traits related to yield, disease resistance, and domestication syndrome.

6. Conservation and Management:
   i. **Evolutionary Rescue:** Conservation efforts increasingly consider evolutionary principles to aid endangered populations. Managed breeding programs sometimes involve the introduction of genetic variation to enhance adaptive potential.
   ii. **Resilience and Restoration:** Understanding evolutionary processes helps in designing ecosystem restoration projects that consider the adaptive potential of species within changing environments.
7. Ethical and Societal Implications:
   i. **Ethical Dilemmas:** The Anthropocene raises ethical questions about human intervention in natural processes, including genetic modification and assisted evolution. Balancing conservation needs with ethical considerations is a significant challenge.
   ii. **Public Awareness:** Educating the public about evolutionary principles in the Anthropocene fosters understanding of the interconnectedness of all life forms and the need for responsible stewardship of the environment.

In short, evolution in the Anthropocene is a multifaceted phenomenon, marked by both challenges and opportunities. Understanding the mechanisms of evolution within human-altered environments is crucial for conservation, management, and ethical decision-making, ensuring the continued adaptability and resilience of Earth's biodiversity in the face of ongoing global changes.

12.3. EMERGING AREAS OF RESEARCH

Evolutionary biology, as a field, continues to evolve itself, constantly embracing new methodologies, technologies, and concepts. Here's an overview of some of the emerging areas of research in evolution:

1. Evolutionary Developmental Biology (Evo-Devo):
   i. **Genetic Basis of Morphological Evolution:** Understanding how changes in developmental genes and regulatory networks lead to morphological diversity across species.
   ii. **Evolution of Novel Structures:** Investigating the genetic

mechanisms underlying the evolution of novel structures and traits during development.

2. Eco-Evolutionary Dynamics:
   i. **Feedback Between Ecology and Evolution:** Studying how ecological interactions drive evolutionary changes in species and how evolutionary changes, in turn, influence ecological dynamics.
   ii. **Rapid Evolution in Changing Environments:** Examining how species adapt to rapidly changing environments, including those influenced by human activities.
3. Microbiome and Evolution:
   i. **Co-evolution with Microorganisms:** Investigating how host organisms and their microbiota co-evolve, influencing each other's evolutionary trajectories.
   ii. **Microbial-Mediated Evolution:** Understanding how microorganisms drive evolutionary changes in host species through processes like symbiosis and pathogenicity.
4. Epigenetics and Evolution:
   i. **Epigenetic Inheritance:** Studying the role of epigenetic modifications in heritable changes and how they influence evolutionary processes.
   ii. **Environmental Epigenetics:** Exploring how environmental factors can induce epigenetic changes, potentially leading to evolutionary adaptations.
5. Evolutionary Systems Biology:
   i. **Network Evolution:** Analyzing the evolution of biological networks, including protein-protein interaction networks and metabolic pathways, to understand their adaptability and robustness.
   ii. **Evolutionary Constraints:** Investigating the constraints imposed by historical evolutionary processes on the present and future evolution of biological systems.
6. Evolutionary Genomics and Big Data:
   i. **Genomic Adaptation:** Identifying genomic signatures of adaptation in natural populations, including the role of non-coding regions and structural variations.
   ii. **Big Data Analytics:** Utilizing big data analytics and machine

learning to analyze large-scale genomic and phenotypic data, uncovering patterns and evolutionary relationships.

7. Evolution of Cooperation and Altruism:
   i. **Evolution of Social Behavior:** Exploring the evolutionary origins of social behaviors, cooperation, altruism, and their genetic and ecological determinants.
   ii. **Game Theory in Evolution:** Applying game theory models to understand the evolution of cooperation and conflict in various biological contexts.
8. Evolution and Conservation:
   i. **Conservation Genomics:** Integrating genomic approaches into conservation biology to understand population dynamics, genetic diversity, and adaptation in endangered species.
   ii. **Resilience in Changing Environments:** Investigating the evolutionary resilience of natural populations and ecosystems facing environmental stressors, including climate change and habitat loss.
9. Evolutionary Ethics and Philosophy:
   i. **Ethical Implications of Evolutionary Research:** Exploring ethical considerations related to evolutionary biology, including issues surrounding genetic engineering, cloning, and ecological preservation.
   ii. **Philosophical Perspectives:** Examining philosophical questions related to the nature of evolution, species concepts, and the implications of evolutionary theory for our understanding of life and humanity.
10. Evolutionary Medicine:
   i. **Evolutionary Basis of Diseases:** Investigating diseases from an evolutionary perspective, understanding how our evolutionary history influences susceptibility, prevalence, and treatment responses.
   ii. **Evolutionary Psychiatry:** Applying evolutionary principles to understand mental health disorders and human cognition, examining how evolutionary pressures shaped our psychological traits.

In conclusion, these emerging areas of research highlight the interdisciplinary nature of evolutionary biology. By integrating genomics, ecology, developmental biology, and ethics, researchers continue to unravel the complexities of evolution, leading to profound insights into the origins and diversity of life on Earth. These evolving fields promise exciting discoveries and applications, shaping our understanding of evolution in the 21st century.

12.4. THE RELEVANCE OF EVOLUTION IN THE 21ST CENTURY

In the 21st century, evolution remains a cornerstone of biological understanding with far-reaching implications across various disciplines. Here's a detailed overview of the relevance of evolution in contemporary times:

1. Understanding Biodiversity:

   i. **Conservation Biology:** Evolutionary principles inform conservation strategies, helping to preserve biodiversity by understanding the genetic diversity within and between species.

   ii. **Ecosystem Functioning:** Studying evolutionary relationships enhances our comprehension of ecosystems, aiding in their conservation and management.

2. Human Health:

   i. **Drug Development:** Evolutionary insights are crucial in combating evolving pathogens, aiding in the development of vaccines and antibiotics to counteract diseases like influenza and antibiotic-resistant bacteria.

   ii. **Cancer Research:** Evolutionary principles help in understanding cancer progression, resistance mechanisms, and the development of targeted therapies.

3. Agriculture and Food Security:

   i. **Crop Improvement:** Evolutionary knowledge guides crop breeding efforts, enhancing crop resilience to diseases, pests, and changing environmental conditions, crucial for global food security.

   ii. **Pest Management:** Understanding the evolutionary dynamics of pests' aids in developing sustainable pest management strategies, reducing crop damage and chemical pesticide use.

4. Biotechnology and Genetic Engineering:
    i. **Gene Editing:** Evolutionary understanding is essential in ethical gene-editing practices like CRISPR-Cas9, ensuring responsible manipulation of genetic material.
    ii. **Synthetic Biology:** Evolutionary concepts inform the design of synthetic organisms and ecosystems, ensuring their compatibility with existing natural systems.
5. Public Health:
    i. **Understanding Disease Spread:** Evolutionary epidemiology helps predict the spread of diseases, offering insights into disease control strategies, particularly in the face of emerging infectious diseases.
    ii. **Vaccine Design:** Evolutionary analysis aids in developing vaccines that can adapt to changing strains of pathogens, ensuring long-term effectiveness.
6. Environmental Science and Climate Change:
    i. **Adaptation to Climate Change:** Evolutionary processes enable species to adapt to changing climates, offering insights into how natural populations might respond to ongoing climate change.
    ii. **Biodiversity Loss:** Evolutionary perspectives highlight the potential consequences of losing unique genetic diversity due to habitat destruction and climate change.
7. Understanding Human Evolution:
    i. **Genetic Anthropology:** Evolutionary genetics traces human migration patterns, origins, and interbreeding events, shaping our understanding of human evolutionary history.
    ii. **Health Disparities:** Evolutionary insights help elucidate genetic factors contributing to health disparities among different populations, informing public health policies and interventions.
8. Education and Public Understanding:
    i. **Promoting Scientific Literacy:** Teaching evolution fosters critical thinking, scientific reasoning, and a deep understanding of the interconnectedness of life, contributing to a scientifically literate society.
    ii. **Countering Misinformation:** Evolutionary education counters misconceptions, promoting evidence-based understanding in the

face of misinformation or anti-scientific beliefs.

9. Ethical and Philosophical Considerations:
   - i. **Ethical Dilemmas:** Evolutionary insights raise ethical questions regarding cloning, gene editing, and the conservation of endangered species, stimulating philosophical and ethical discussions.
   - ii. **Philosophical Relevance:** Evolutionary theory continues to influence philosophical discussions on topics like human nature, morality, and the nature of life itself.

In summary, evolution's relevance in the 21st century extends far beyond biology classrooms. It plays a central role in addressing global challenges, from public health crises to environmental sustainability. Its multidisciplinary applications underscore the critical importance of evolutionary understanding in shaping a sustainable and informed future for humanity.

REFERENCES

1. Eldredge, N., (2015). *Eternal Ephemera.*
2. Fedonkin, M. A., et al., (2008). *The Rise of Animals.*
3. Gee, H., (2013). *The Accidental Species: Misunderstandings of Human Evolution.*
4. Hodder, (2018). *Where Are We Heading? The Evolution of Humans and Things.*
5. Jablonka, E., & Lamb, M. J., (2005). *Evolution in Four Dimensions.*
6. Larson, E. J., (2004). *Evolution.*
7. Marks, J., (2021). *Why Are There Still Creationists? Human Evolution and the Ancestors.*
8. Mayr, E., (2001). *What Evolution Is?*
9. Niklas, K. J., (2016). *Plant Evolution: An Introduction to the History of Life.*
10. Palmer, D., (2009). *Evolution.*
11. Pigliucci, M., & Müller, G. B., (2010). *Evolution, the Extended Synthesis.*
12. Prothero, D. R., (2017). *Evolution: What the Fossils Say and Why It Matters* (2nd edn.).
13. Rogers, R., (2011). *The Evidence for Evolution.*
14. Ruse, M., (2005). *The Evolution-Creation Struggle.*
15. Russell, D. A., (2009). *Islands in the Cosmos: The Evolution of Life on Land.*
16. Scott, E. C., (2004). *Evolution vs. Creationism.*
17. Stott, R., (2012). *Darwin's Ghosts: The Secret History of Evolution.*

CHAPTER 13

Conclusion and Looking Ahead

CONTENTS

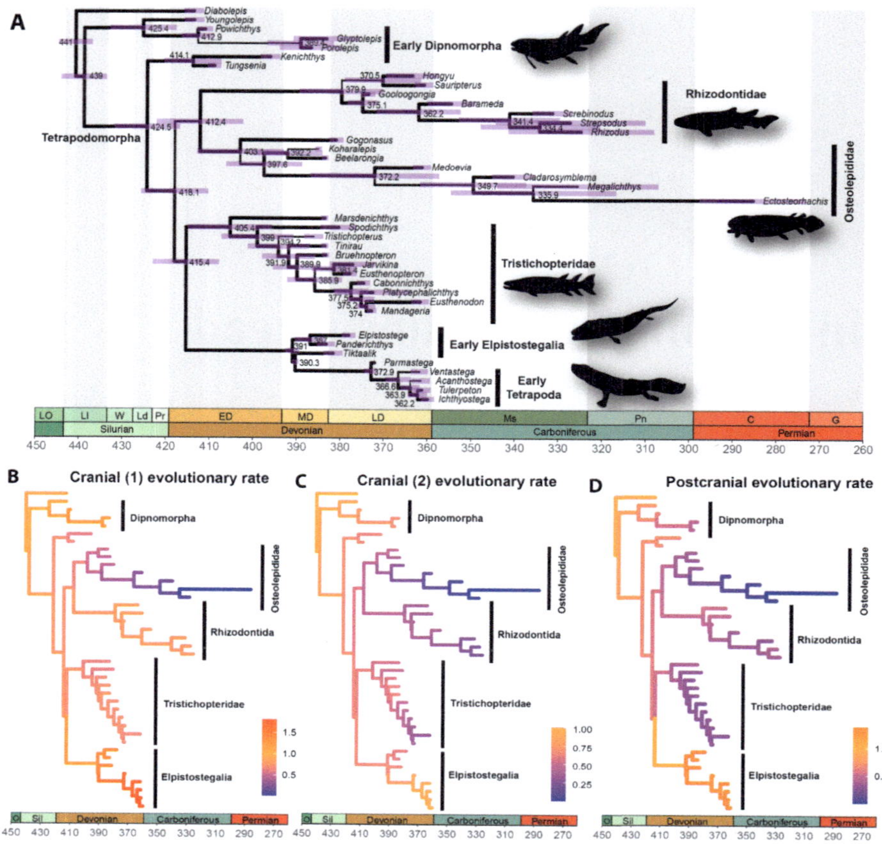

Figure 13.1. Looking ahead in evolutionary biology.

Source: López-Antoñanzas, R., Mitchell, J., Simões, T. R., Condamine, F. L., Aguilée, R., Peláez-Campomanes, P., Renaud, S., Rolland, J., & Donoghue, P. C. J., (2022). Integrative phylogenetics: tools for paleontologists to explore the tree of life. Biology, 11(8), 1185. https://doi.org/10.3390/biology11081185; Creative Commons License.

13.1. THE ONGOING IMPACT OF EVOLUTIONARY THEORY

Evolutionary theory stands as one of the most transformative scientific ideas in human history, fundamentally shaping our understanding of life on Earth. From its origins in the 19th century to its diverse applications in the 21st century, evolutionary theory has influenced nearly every branch of science, from genetics and ecology to medicine and anthropology. This enduring impact can be attributed to the elegance of its core principles—descent with modification, natural selection, genetic variation, and speciation—which have provided a unifying framework for interpreting the vast diversity of life forms. Through rigorous scientific inquiry and technological advancements, researchers have continued to refine and expand these foundational concepts, revealing intricate details about the mechanisms driving evolution.

13.1.1. Looking Ahead: The Future of Evolutionary Theory

As we move forward, evolutionary theory remains at the forefront of scientific exploration, leading to exciting avenues of research and discovery. Here are key aspects shaping the future of evolutionary theory:

1. **Genomic Revolution:** Advances in genomics, coupled with big data analytics, are unraveling the genetic basis of evolution with unprecedented precision. Understanding the intricacies of genomes across diverse species will deepen our insights into adaptation, speciation, and evolutionary history.
2. **Eco-Evolutionary Dynamics:** Integrating ecology and evolution, researchers are exploring how species' interactions with their environments drive evolutionary changes. This interdisciplinary approach is illuminating the reciprocal impacts of ecological shifts and evolutionary adaptations.
3. **Evolutionary Developmental Biology (Evo-Devo):** Investigating the links between genetics and morphology, Evo-Devo continues to uncover the regulatory mechanisms shaping the diverse forms seen in nature. Understanding how genes control development provides profound insights into evolutionary innovations.
4. **Human Evolution and Cultural Evolution:** Studying the evolution of Homo sapiens, both biologically and culturally, sheds light on our shared ancestry and the factors shaping human societies. Cultural evolution, in particular, is becoming a central focus, exploring the transmission and adaptation of cultural traits over generations.
5. **Evolution in the Anthropocene:** In the face of human-induced environmental changes, studying how species adapt, evolve, or face extinction is crucial. Evolutionary insights guide conservation efforts, help predict species' responses to climate change, and inform sustainable practices.

6. **Ethical and Societal Implications:** Evolutionary theory raises profound ethical questions related to genetic engineering, biodiversity conservation, and our responsibility toward other species. Addressing these ethical challenges will be essential in the future discourse of evolutionary science.
7. **Education and Outreach:** Fostering a deeper understanding of evolutionary principles among the public is vital. Promoting scientific literacy and countering misinformation are essential tasks, ensuring that evolutionary science remains accessible and relevant to society.

In the coming years, as evolutionary theory continues to advance, its impact will reach far beyond academic circles. It will guide conservation efforts, improve human health, enhance agricultural practices, and deepen our understanding of life's intricate complexities. Through ongoing research, ethical considerations, and educational initiatives, the legacy of evolutionary theory will not only endure but also flourish, shaping the scientific landscape and our understanding of the natural world for generations to come.

13.2. FUTURE CHALLENGES AND DISCOVERIES

As we peer into the future of evolutionary biology, several challenges and exciting discoveries lie on the horizon, shaping the landscape of this field in profound ways.

1. Unraveling Complexity:
   - i. **Challenges:** Exploring the complexities of gene-environment interactions, epigenetics, and multi-level selection poses intricate challenges. Deciphering the intricacies of how genes and environments interact to shape traits will require innovative methods and interdisciplinary collaboration.
   - ii. **Discoveries:** Advancements in computational biology, combined with high-throughput data analysis, will likely reveal hidden patterns in complex biological systems. Understanding the interplay between genetics and the environment will lead to breakthroughs in areas like personalized medicine and ecological prediction.
2. Integrating Evolution and Development:
   - i. **Challenges:** Integrating evolutionary and developmental biology (Evo-Devo) presents challenges in bridging molecular

genetics and macroevolutionary patterns. Understanding how developmental processes influence long-term evolutionary changes requires in-depth molecular and morphological studies.

ii. **Discoveries:** Cutting-edge techniques like single-cell genomics and advanced imaging methods will allow scientists to explore the evolutionary significance of developmental processes at unprecedented resolutions. Discoveries in this area could revolutionize our understanding of evolutionary innovation and morphological diversity.

3. Human Evolutionary Genomics:

i. **Challenges:** Tracing the intricacies of human evolution, especially understanding archaic hominin admixture and migration patterns, remains a challenge. Unraveling the genetic basis of uniquely human traits and diseases in the context of our evolutionary history presents complex puzzles.

ii. **Discoveries:** Ongoing advances in ancient DNA sequencing and population genomics will likely yield detailed insights into human evolutionary history. Discoveries might include identifying previously unknown hominin species, understanding the genetic basis of human cognition, and elucidating the evolutionary origins of diseases like cancer and neurodegenerative disorders.

4. Evolution in Changing Environments:

i. **Challenges:** Studying evolution in rapidly changing environments, such as urban landscapes and under climate change, presents challenges due to the pace of adaptation required. Predicting which species will thrive, adapt, or face extinction in these novel environments is a complex task.

ii. **Discoveries:** Genomic studies of urban-adapted species and long-term ecological monitoring will provide insights into rapid evolutionary changes. Understanding the mechanisms enabling species to adapt to human-altered environments will have significant implications for conservation strategies and ecosystem management.

5. Ethical and Societal Challenges:

i. **Challenges:** The ethical dilemmas associated with genetic engineering, gene editing, and de-extinction efforts demand careful consideration. Addressing the ethical implications of interventions in natural evolutionary processes remains a challenge.

ii. **Discoveries:** Ethical frameworks and guidelines will likely evolve alongside technological advancements. Discoveries might include novel ethical models for conservation efforts, ensuring the preservation of biodiversity while respecting the autonomy of ecosystems and species.

6. Advancing Public Understanding:

i. **Challenges:** Countering misinformation and fostering public understanding of evolutionary concepts remain challenges, particularly in regions with contentious sociopolitical climates.

ii. **Discoveries:** Innovative educational approaches, multimedia tools, and interactive platforms will likely emerge to enhance public engagement. Discoveries in effective science communication methods will bridge the gap between scientific knowledge and public understanding, promoting informed discussions about evolutionary concepts.

In the face of these challenges, the discoveries awaiting evolutionary biology are bound to be transformative. With interdisciplinary collaboration, technological innovations, and a commitment to ethical practices, the future of evolutionary biology holds the promise of unraveling some of life's most profound mysteries while addressing critical challenges facing our planet and its biodiversity.

REFERENCES

1. Alison, J., (2001). *Lucy's Legacy: Sex and Intelligence in Human Evolution*.
2. Ann, G., (2007). *The First Human: The Race to Discover Our Earliest Ancestors*.
3. Arik, K., (2020). *The Zoologist's Guide to the Galaxy – What Animals on Earth Reveal about Aliens – and Ourselves*.
4. Bernard, W., (2006). *Human Evolution: A Very Short Introduction*.
5. Bill, N., (2014). *Undeniable: Evolution and the Science of Creation*.
6. Brian, C., & Deborah, C., (2003). *Evolution: A Very Short Introduction*.
7. Cameron, M. S., & Charles, S., (2006). *The Top 10 Myths About Evolution*.
8. Carl, S., (1977). *The Dragons of Eden: Speculations on the Evolution of Human Intelligence*.
9. Carl, Z., (2001). *Evolution: The Triumph of an Idea*.
10. Carl, Z., (2005). *Smithsonian Intimate Guide to Human Origins*.
11. Charles, D., (1859). *The Origin of Species*.
12. Charles, D., (1871). *The Descent of Man*.
13. Daniel, D., (1995). *Darwin's Dangerous Idea*.
14. Daniel, D., (2003). *Freedom Evolves*.
15. Daniel, J. F., (2007). *Relics of Eden: The Powerful Evidence of Evolution in Human DNA*.
16. Daniel, L., (2008). *The World in Six Songs: How the Musical Brain Created Human Nature*.
17. David, S. J., (1901). *The Blood of the Nation: A Study in the Decay of Races by the Survival of the Unfit*.
18. David, S. W., (2007). *Evolution for Everyone: How Darwin's Theory Can Change the Way We Think About Our Lives*.
19. David, S. W., (2019). *This View of Life: Completing the Darwinian Revolution*
20. Donald, J., (2006). *From Lucy to Language: Revised, Updated, and Expanded*.
21. Donald, P., (2006). *After the Dinosaurs: The Age of Mammals*.
22. Donald, P., (2007). *Evolution: What the Fossils Say and Why It Matters*.

23. Ernst, M., (2002). *What Evolution Is?*
24. Ernst, M., (2007). *One Long Argument: Charles Darwin and the Genesis of Modern Evolutionary Thought.*
25. Gabriel, E. L., (1961). *The Evolution of Man: A Brief Introduction to Physical Anthropology.*
26. Jared, D., (1992). *The Third Chimpanzee: The Evolution and Future of the Human Animal.*
27. Jay, H., (2011). *Evolution: The Story of Life on Earth.*
28. Jean-Baptiste, D. P., (2011). *Evolution.*
29. Jerry, C., (2009). *Why Evolution Is True.*
30. John, M. S., & Eörs, S., (1997). *The Major Transitions in Evolution.*
31. John, M. S., & Eörs, S., (1999). *The Origins of Life: From the Birth of Life to the Origin of Language.*
32. John, M. S., (1958). *The Theory of Evolution.*
33. John, M. S., (1972). *On Evolution.*
34. John, M. S., (1978). *The Evolution of Sex.*
35. John, M. S., (1982). *Evolution and the Theory of Games.*
36. John, M. S., (1989). *Evolutionary Genetics.*
37. John, S., & Dorion, S., (2002). *Up from Dragons: The Evolution of Human Intelligence.*
38. Jonathan, W., (1994). *The Beak of the Finch: A Story of Evolution in Our Time.*
39. Joseph, J., (2006). *Who Asked the First Question? The Origins of Human Choral Singing, Intelligence, Language, and Speech.*
40. Joseph, J., (2011). *Why do People Sing? Music in Human Evolution.*
41. Julian, H., (1942). *Evolution: The Modern Synthesis.*
42. Kenneth, R. M., (2000). *Finding Darwin's God: A Scientist's Search for Common Ground Between God and Evolution.*
43. Kenneth, R. M., (2008). *Only a Theory: Evolution and the Battle for America's Soul.*
44. Marc, H., (2006). *Moral Minds.*
45. Marc, W. K., & John, C. G., (2006). *The Plausibility of Life: Resolving Darwin's Dilemma.*

46. Matt, R., (1994). *The Red Queen: Sex and the Evolution of Human Nature.*
47. Matt, R., (1997). *The Origins of Virtue.*
48. Matt, R., (1999). *Genome: The Autobiography of a Species in 23 Chapters.*
49. Matt, R., (2003). *Nature Via Nurture: Genes, Experience, and What Makes Us Human.*
50. Matteo, C., (2008). *The Selfish Cell: An Evolutionary Defeat.*
51. Michael, R., (1998). *Taking Darwin Seriously: A Naturalistic Approach to Philosophy.*
52. Michael, R., (1999). *The Darwinian Revolution: Science Red in Tooth and Claw.*
53. Michael, R., (2001). *Mystery of Mysteries: Is Evolution a Social Construction?*
54. Michael, R., (2001). *The Evolution Wars: A Guide to the Debates.*
55. Michael, R., (2006). *Darwinism and Its Discontents.*
56. Michael, R., (2006). *Evolution-Creation Struggle.*
57. National Academies of Science, (2008). *Science, Evolution, and Creationism.*
58. Neil, S., (2008). *Your Inner Fish: A Journey into the 3.5-Billion-Year History of the Human Body.*
59. Nicholas, W., (2006). *Before the Dawn: Recovering the Lost History of Our Ancestors.*
60. Nick, L., (2010). *Life Ascending: The 10 Great Inventions of Evolution.*
61. Niles, E., (2001). *The Triumph of Evolution and the Failure of Creationism.*
62. Paul, M. B., & Joanne, S., (2009). *Death from a Distance and the Birth of a Humane Universe: Human Evolution, Behavior, History, and Your Future.*
63. Peter, K., (1902). *Mutual Aid: A Factor of Evolution.*
64. *Reprinted as The Agile Gene: How Nature Turns on Nurture.*
65. Richard, D., (1976). *The Selfish Gene.*
66. Richard, D., (1982). *The Extended Phenotype.*
67. Richard, D., (1986). *The Blind Watchmaker.*
68. Richard, D., (1995). *River Out of Eden.*

69. Richard, D., (1996). *Climbing Mount Improbable.*
70. Richard, D., (2004). *The Ancestor's Tale.*
71. Richard, D., (2009). *The Greatest Show on Earth: The Evidence for Evolution.*
72. Sean, B. C., (2005). *Endless Forms Most Beautiful: The New Science of Evo Devo.*
73. Sean, B. C., (2006). *The Making of the Fittest: DNA and the Ultimate Forensic Record of Evolution.*
74. Shaun, J., (2017). *Re-Thinking What it Means We Evolved: A New Framework for Universal Moral Values.*
75. Shaun, J., (2020). *Are You Wonderful? Good Science Says, Yes: How to Tell Good Science from Bad.*
76. Stephen, J. G., (1977). *Ever Since Darwin.*
77. Stephen, J. G., (1989). *Wonderful Life: The Burgess Shale and the Nature of History.*
78. Stephen, J. G., (1996). *Full House: The Spread of Excellence from Plato to Darwin.*
79. Steve, J., (1995). *The Language of the Genes.*
80. Steven, P., (1997). *How the Mind Works*?
81. Theodosius, D., (1937/1941/1951). *Genetics and the Origin of Species* (2nd & 3rd edns.).
82. Thomas, H. H., (1863). *Evidence as to Man's Place in Nature.*

Appendix 1: The Evolution of Life Timeline

Overview

Hadean eon

3800 Ma and earlier.

Date	Event
4567.17 Ma	The planet Earth forms from the accretion disc revolving around the young Sun.
4533 Ma	The planet Earth and the planet Theia collide, sending countless moonlets into orbit around the young Earth. These moonlets eventually coalesce to form the Moon. The gravitational pull of the new Moon stabilizes the Earth's fluctuating axis of rotation and sets up the conditions for the formation of life.
4100 Ma	The surface of the Earth cools enough for the crust to solidify. The atmosphere and the oceans form.

Between 4500 and 2500 Ma	The earliest life appears, possibly derived from self-reproducing RNA molecules. The replication of these organisms requires resources like energy, space, and smaller building blocks, which soon become limited, resulting in competition. Natural selection favors those molecules which are more efficient at replication. DNA molecules then take over as the main replicators. They soon develop inside enclosing membranes which provide a stable physical and chemical environment conducive to their replication: proto cells.
3900 Ma	Late Heavy Bombardment: peak rate of impact events upon the inner planets by meteors. This constant disturbance probably obliterated any life that had already evolved, as the oceans boiled away completely; conversely, life may have been transported to Earth by a meteor.
Somewhere between 3900–2500 Ma	Cells resembling prokaryotes appear. These first organisms are chemoautotrophs: they use carbon dioxide as a carbon source and oxidize inorganic materials to extract energy. Later, prokaryotes evolve glycolysis, a set of chemical reactions that free the energy of organic molecules such as glucose. Glycolysis generates ATP molecules as short-term energy currency, and ATP continues to be used in almost all organisms, unchanged, to this day.

Archean Eon

3800 Ma–2500 Ma

Date	Event
3500 Ma	Lifetime of the last universal ancestor; the split between the bacteria and the archaea occurs. Bacteria develop primitive forms of photosynthesis which at first do not produce oxygen. These organisms generate ATP by exploiting a proton gradient, a mechanism still used in virtually all organisms.

3000 Ma	Photosynthesizing cyanobacteria evolve; they use water as a reducing agent, thereby producing oxygen as a waste product. The oxygen initially oxidizes dissolved iron in the oceans, creating iron ore. The oxygen concentration in the atmosphere subsequently rises, acting as a poison for many bacteria. The moon is still very close to the earth and causes tides 1000 feet high. The earth is continually wracked by hurricane force winds. These extreme mixing influences are thought to stimulate evolutionary processes. (See Oxygen Catastrophe)

Proterozoic Eon

2500 Ma–542 Ma

Date	Event
By 2100 Ma	Eukaryotic cells appear. Eukaryotes contain membrane-bound organelles with diverse functions, probably derived prokaryotes engulfing each other via phagocytosis.
By 1200 Ma	Sexual reproduction evolves, increasing the rate of evolution.
1200 Ma	Simple multicellular organisms evolve, mostly consisting of cell colonies of limited complexity.
850–630 Ma	A global glaciation may have reduced the diversity of life. Opinion is divided on whether it increased or decreased the rate of evolution.
580–542 Ma	The Ediacaran biota represent the first large, complex multicellular organisms – although their affinities remain a subject of debate.
580–500 Ma	Most modern groups begin to appear in the fossil record during the Cambrian explosion.
Around 540 Ma	The accumulation of atmospheric oxygen allows the formation of an ozone layer. This blocks ultraviolet radiation, permitting the colonization of the land.

Phanerozoic Eon

542 Ma–Present

The Phanerozoic eon, literally the "period of well-displayed life," marks the appearance in the fossil record of abundant, shell-forming, and/or trace-making organisms. It is subdivided into three eras, the Paleozoic, Mesozoic, and Cenozoic, which are divided by major mass extinctions.

Paleozoic Era

542 Ma–251.0 Ma

Date	Event
530 Ma	The first known footprints on land date to 530 Ma, indicating that early animal explorations may have predated the development of terrestrial plants.
475 Ma	The first primitive plants move onto land, having evolved from green algae living along the edges of lakes. They are accompanied by fungi, which may have aided the colonization of land through symbiosis.
363 Ma	By the start of the Carboniferous period, the Earth begins to be recognizable. Insects roamed the land and would soon take to the skies; sharks predated the oceans, and vegetation covered the land, with seed-bearing plants and forests soon to flourish. Four-limbed tetrapods gradually gain adaptions which will help them occupy a terrestrial life-habit.
251.4 Ma	The Permian-Triassic extinction event eliminates over 95% of species. This "clearing of the slate" may have led to an ensuing diversification.

Mesozoic Era

Date	Event
From 251.4 Ma	The Mesozoic Marine Revolution begins increasingly well-adapted and diverse predators pressurize sessile marine groups; the "balance of power" in the oceans shifts dramatically as some groups of prey adapt more rapidly and effectively than others.

220 Ma	Gymnosperm forests dominate the land; herbivores grow to huge sizes in order to accommodate the large guts necessary to digest the nutrient-poor plants
200 Ma	The first accepted evidence for viruses – the group Geminiviridae at least exists. Viruses are still poorly understood and may have arisen before "life" itself or may be a more recent phenomenon.
130 Ma	The rise of the Angiosperms: These flowering plants boast structures that attract insects and other animals to spread pollen. This innovation causes a major burst of animal evolution through co-evolution.

Cenozoic Era

65.5 Ma–Present

Date	Event
65.5 Ma	The Cretaceous–Tertiary extinction event eradicates about half of all animal species, including all non-avian dinosaurs.
35 Ma	Grasses evolve from among the angiosperms; grassland dominates many terrestrial ecosystems.
14,000 years ago	The term Anthropocene has been used to describe the period of time during which Man has had a major impact on the planet and its inhabitants. Its beginning is marked by the megafaunal extinction in the Americas which signify the onset of the Holocene extinction event. Fierce debate rages about the influence of man in the initiation of this extinction, but no one can deny that humanity is contributing to its propagation.
Present day	With a human population approaching 6.67 billion, the impact of humanity is felt in all corners of the globe. Overfishing, anthropogenic climate change, industrialization, intensive agriculture, clearance of rain forests and other activities contribute to a dramatically rising extinction rate. At current rates, humanity will have eradicated one-half of Earth's biodiversity over the next 100 years.

Detailed

4.5 Billion Years Ago:

1. Earth Forms:

i. Soon afterward it is struck by a Mars-sized body dubbed Theia, which vaporizes the surface and blasts it into space. This ejected material condenses to form the Moon.

ii. Massive bodies continue to strike the Earth, at a declining rate, for the next 1.5 billion years, ending about 3 billion years ago. The impacts reshape the planet's surface and may help drive the onset of plate tectonics.

iii. However, the idea that there is an especially intense "Late Heavy Bombardment" between about 4 and 3.8 billion years ago, which obliterates any life, no longer seems to be true. This means life may have begun in the first few 100 million years of Earth's existence.

4.4 Billion Years Ago:

i. By this time, Earth has cooled down considerably and has substantial amounts of water. The origin of this water is mysterious. It may have been contained in the rocks that accreted to form the planet, in which case Earth was born wet.

ii. Alternatively, the planet may have been dry at first, and received water from impactors like comets and asteroids. Either way, the early presence of water also points to an early origin of life.

4.29 Billion Years Ago:

i. Chemical traces in rocks from Canada may represent evidence of life from a deep-sea vent, but the results are contested.

ii. When and how life begins remains highly uncertain. There are multiple competing hypotheses for the nature of the first life, for instance that it was based on RNA rather than DNA.

iii. There are also many proposed settings, such as undersea alkaline vents and geothermal ponds on land. The question of life's origin fuses a range of scientific disciplines from chemistry to ecology.

4.2 Billion Years Ago:

i. The first large oceans may form at this time. These primordial seas may be much deeper than today, leaving little or no exposed land.

4.1 Billion Years Ago:

i. Flecks of carbon trapped inside grains of zircon suggest that life exists at this time, but the evidence is limited.

3.9 Billion Years Ago:

i. The Last Universal Common Ancestor (LUCA), the species from which everything alive today is descended, may live at this time according to genetic evidence. Attempts to reconstruct its genome suggest it lives in a volcanic or geothermal setting. LUCA seems to be a relatively complex microorganism, indicating that the origin of life occurred significantly earlier.

ii. LUCA gives rise to two main groups of life, both single-celled: bacteria and archaea. How and when this split happens is still uncertain.

3.7 Billion Years Ago:

i. Rocks in Greenland from this time contain mysterious structures that may be fossilized microorganisms or may just be distorted rocks.

ii. The first large land masses may emerge from the ocean around this time, following a change in the behavior of tectonic plates.

3.5 Billion Years Ago:

i. The oldest fossils of single-celled organisms, from Pilbara in Western Australia, date from this time.

ii. They may live in freshwater hot springs in a volcanic region on land. The ecosystem is complex and thriving, suggesting life is already well-established.

3.46 Billion Years Ago:

i. Some single-celled organisms may be feeding on methane by this time.

3.4 Billion Years Ago:

i. Some bacteria are performing photosynthesis: they take in sunlight, carbon dioxide, and obtain energy. However, this is not photosynthesis as we know it today, because the bacteria do not release oxygen as a waste product. This anoxygenic photosynthesis remains common for a billion years.

3.2 Billion Years Ago:

i. Fossil microorganisms preserved in rocks from South Africa offer undisputed evidence of life on land.

3 Billion Years Ago:

i. Viruses are present by this time, but they may be as old as life itself, and have influenced the course of its evolution.

2.5 to 2.2 Billion Years Ago:

i. The Great Oxidation Event. Some cyanobacteria evolve a new form of photosynthesis that releases oxygen. This toxic waste starts to build up in the seas and atmosphere – though concentrations remain below modern levels for over a billion years.
ii. Dissolved oxygen makes the iron in the oceans "rust" and sink to the seafloor, forming striking banded iron formations.
iii. Oxygenic photosynthesis may well have evolved earlier, but the resulting oxygen was initially confined to small "oases."
iv. Once oxygen becomes widespread, it may have caused a mass extinction among microbes that are unable to cope with it. It also drives evolutionary innovations. Today almost all animals breathe it, and it may be behind the origin of circadian clocks.

2.3 Billion Years Ago:

i. The earth freezes over in what may have been the first "snowball Earth," possibly as a result of a lack of volcanic activity. When the ice eventually melts, it indirectly leads to more oxygen being released into the atmosphere.

2.1 Billion Years Ago:

i. Burrow-like structures in rocks from Gabon suggest simple multicellular organisms have evolved and are moving, but this is disputed.

2 Billion Years Ago:

i. Eukaryotic cells – cells with internal "organs" (known as organelles) – come into being. One key organelle is the nucleus: the control center of the cell, in which the genes are stored in the form of DNA.

ii. Fossils resembling eukaryotic cells are known from 2 billion years ago, and there are disputed claims of older fossils.

iii. Eukaryotic cells evolve when one simple cell engulfed another, and the two lived together, more, or less amicably – an example of "endosymbiosis." The engulfed bacteria eventually become mitochondria, which provide eukaryotic cells with energy. The last common ancestor of all eukaryotic cells had mitochondria – and had also developed sexual reproduction.

iv. The origin of eukaryotes is a turning point in evolution. It is not clear why it happened, but microbes called Asgard archaea may offer clues. Of all known archaea, they are the most closely related to eukaryotes, so ancient archaea similar to them may have taken part in the endosymbiosis.

v. Later, some eukaryotic cells engulf photosynthetic bacteria and form a symbiotic relationship with them. The engulfed bacteria evolve into chloroplasts: the organelles that give green plants their color and allow them to extract energy from sunlight.

1.6 Billion Years Ago:

i. Fossils resembling primitive seaweed suggest some eukaryotes have evolved into large, multicellular organisms. The timing of the origin of multicellularity is uncertain and it may have occurred multiple times.

ii. It is unclear exactly how or why multicellularity arises. One possibility is that single-celled organisms go through a stage similar to that of modern choanoflagellates: single-celled creatures that sometimes form colonies consisting of many individuals. Of all the single-celled organisms known to exist, choanoflagellates are the most closely related to multicellular animals, lending support to this idea.

1.5 Billion Years Ago:

i. The eukaryotes undergo major splits: the ancestors of modern plants, fungi, and animals split into separate lineages, and evolve separately. We do not know in what order the three groups broke with each other.

1 Billion Years Ago:

i. Some simple multicellular organisms appear to have distinct cell types, the first step towards internal organs.

890 Million Years Ago:

i. The first recognizable animals may have evolved by this time, much earlier than was once believed. In particular, fossils indicate that sponges are present.

800 Million Years Ago:

i. Around this time, fungi evolve multicellular body plans.

ii. Meanwhile, the early multicellular animals undergo their first splits. First, they divide into, essentially, the sponges and everything else – the latter being more formally known as the Eumetazoa.

iii. However, some genetic studies indicate that a group called the comb jellies or ctenophores was the first to break away, not the sponges. The debate is ongoing.

iv. Around 20 million years later, a small group called the placozoa breaks away from the rest of the Eumetazoa. Placozoa are thin plate-like creatures about 1 millimeter across and consist of only three layers of cells.

717 to 636 Million Years Ago:

i. The planet freezes over again in another "snowball Earth."

660 Million Years Ago:

i. Chemical traces in rocks suggest sponges are living in the sea.

630 Million Years Ago:

i. Around this time, some animals evolve bilateral symmetry for the first time: that is, they now have a defined top and bottom, as well as a front and back.

ii. Little is known about how this happened. However, small worms called Acoela may be the closest surviving relatives of the first ever bilateral animal.

iii. It seems likely that the first bilateral animal was a kind of worm. Vernanimalculaguizhouena, which dates from around 600 million years ago, may be the earliest bilateral animal found in the fossil record.

600 Million Years Ago:

i. Fossils that may be comb jellies date from this time.

590 Million Years Ago:

i. The Bilateria, those animals with bilateral symmetry, undergo a profound evolutionary split. They divide into the protostomes and deuterostomes.

ii. The deuterostomes eventually include all the vertebrates, plus an outlier group called the Ambulacraria. The protostomes become all the arthropods (insects, spiders, crabs, shrimp, etc.), mollusks, various types of worms, and the microscopic rotifers.

iii. Neither may seem like an obvious "group," but in fact the two can be distinguished by the way their embryos develop. The first hole that the embryo acquires, the blastopore, forms the anus in deuterostomes, but in protostomes it forms the mouth.

580 Million Years Ago:

i. The earliest known fossils of cnidarians, the group that includes jellyfish, sea anemones and corals, date to around this time – though the fossil evidence has been disputed.

575 Million Years Ago:

i. Strange life forms known as the Ediacarans appear around this time and persist for about 33 million years. Some may be early animals.

570 Million Years Ago:

i. The Ambulacraria breaks away from the main group of deuterostomes. This small group eventually becomes the echinoderms (starfish, brittle stars, and their relatives) and two worm-like families called the hemichordates and Xenoturbellida.
ii. Another echinoderm, the sea lily, is thought to be the "missing link" between vertebrates (animals with backbones) and invertebrates (animals without backbones), a split that occurred around this time.

565 Million Years Ago:

i. Fossilized animal trails suggest that some animals are moving under their own power.

558 Million Years Ago:

i. Dickinsonia, an Ediacaran that is thought to be an early animal, lives around this time.

540 Million Years Ago:

i. As the first chordates – animals that have a backbone, or at least a primitive version of it – emerge among the deuterostomes, a surprising cousin branches off.

ii. The sea squirts (tunicates) begin their history as tadpole-like chordates but metamorphose partway through their lives into bottom-dwelling filter feeders that look rather like a bag of seawater anchored to a rock. Their larvae still look like tadpoles today, revealing their close relationship to backboned animals.

iii. Around the same time, the Ediacarans die out, perhaps due to competition from newly evolved species.

539 Million Years Ago:

i. The Cambrian explosion begins, with many new body layouts appearing on the scene – though the seeming rapidity of the appearance of new life forms increasingly looks like an illusion as ever more older fossils come to light.

530 Million Years Ago:

i. The first true vertebrates – animals with a backbone – appear. The exact timing is unclear, and it is difficult to find consensus on which animals are true vertebrates and which are only related to the vertebrates. Possible candidates include Myllokunmingia, known from 518-million-year-old rocks in China.

ii. The early vertebrates have a notochord, protected by a stiff rod of cartilage: they do not have a spine made of bone. The first vertebrate is probably quite like a lamprey, hagfish, or lancelet.

iii. Around the same time, the first clear fossils of trilobites appear. These invertebrates, which look like oversized woodlice and grow to 70 centimeters in length, proliferate in the oceans for the next 200 million years.

520 Million Years Ago:

i. Conodonts, another contender for the title of "earliest vertebrate," appear. They look a little like eels – although whether they are truly vertebrates is still debated.

515 Million Years Ago:

i. Plants begin colonizing the land – an event that may have been enabled by genes obtained from microorganisms.

512 Million Years Ago:

i. A fossilized worm indicates that parasites are now present on Earth, but this lifestyle may be much older.

500 Million Years Ago:

i. Fossil evidence shows that animals were exploring the land at this time. The first animals to do so were probably euthycarcinoids – thought to be related to centipedes and millipedes.

ii. *Nectocaris pteryx* lives around this time. It is thought by some researchers to be the oldest known ancestor of the cephalopods – the group that includes squid – although others are not so sure.

iii. Some comb jellies develop complex nervous systems.

489 Million Years Ago:

i. The Great Ordovician Biodiversification Event begins, leading to a great increase in the diversity of life. Within each of the major groups of animals and plants, many new varieties appear.

450 to 400 Million Years Ago:

i. Fish split into two major groups: the bony fish and cartilaginous fish. The cartilaginous fish, as the name implies, have skeletons made of cartilage rather than the harder bone. Today, they include all the sharks, skates, and rays.

ii. Fish with teeth appear in the fossil record at this time, including Qianodus, thought to have been a cartilaginous fish.

iii. Oxygen reaches its modern level in the atmosphere for the first time – although the exact level will continue to fluctuate.

445–443 Million Years Ago:

i. The Late Ordovician mass extinction, the first of the "big five" extinction events and probably the second most severe in terms of the number of genera that went extinct.

440 Million Years Ago:

i. The bony fish split into their two major groups: the lobe-finned fish with bones in their fleshy fins, and the ray-finned fish. The lobe-finned fish eventually give rise to amphibians, reptiles, birds, and mammals. The ray-finned fish thrive and give rise to most fish species living today.

ii. The common ancestor of lobe-finned and ray-finned fish probably has simple sacs that function as primitive lungs, allowing it to gulp air when oxygen levels in the water fall too low. In ray-finned fish, these sacs evolve into the swim bladder, which is used for controlling buoyancy.

432 Million Years Ago:

i. The oldest unambiguous large fossil of a land plant dates from this time.

425 Million Years Ago:

i. The coelacanths, a group of famous "living fossils" – species that have apparently not changed for millions of years – may have split from the rest of the lobe-finned fish at this time. Coelacanths begin appearing in the fossil record by 409 million years ago.

417 Million Years Ago:

i. The lungfish, another group of legendary living fossils, split from the other lobe-finned fish. Although they are unambiguously fish, complete with gills, lungfish have a pair of relatively sophisticated lungs, which are divided into numerous smaller air sacs to increase their surface area. These allow them to breathe out of water and thus to survive when the ponds they live in dry out.

410 Million Years Ago:

i. For the first time, plants with elaborate root systems that stabilize the soil leave their mark in the fossil record.

400 Million Years Ago:

i. Insects may start appearing in the fossil record around this time – although some researchers now suspect the fossils are misidentified millipedes.

ii. Some plants evolve woody stems.

397 Million Years Ago:

i. Fossilized footprints are left by four-legged animals, or tetrapods, for the first time. But the fossils are controversial: they seem to predate, by many millions of years, some of the fish-like animals thought to have evolved into tetrapods, and so some researchers question whether the fossils really are ancient footprints.

385 Million Years Ago:

i. The oldest fossilized tree dates from this period.

375 Million Years Ago:

i. Tiktaalik, an intermediate between fish and four-legged land animals, lives around this time. The fleshy fins of its lungfish ancestors are evolving into limbs, and unambiguous evidence that tetrapods have made it onto land soon follows.

ii. The tetrapods eventually give rise to all amphibians, reptiles, birds, and mammals.

iii. But not all Tiktaalik-like animals move on to land. One – Qikiqtania – evolves back into an aquatic swimmer.

372 to 359 Million Years Ago:

i. The Late Devonian mass extinction occurs. It is the second of the big five. This one may be a long slow decline over tens of millions of years, not an abrupt event.

350 Million Years Ago:

i. The beetles, one of the most successful animal groups, begin to proliferate.

340 Million Years Ago:

i. The first major split occurs in the tetrapods, with the amphibians branching off from the others.

310 Million Years Ago:

i. Within the remaining tetrapods, the sauropsids and synapsids split from one another. The sauropsids include all the modern reptiles, plus the dinosaurs and birds. The first synapsids are also reptiles but have distinctive jaws. They are sometimes called "mammal-like reptiles," and eventually some of them evolve into the mammals.

320 to 250 Million Years Ago:

i. Synapsid animals known informally as the "pelycosaurs" dominate the land. The most famous example is Dimetrodon, a large predatory "reptile" with a sail on its back. Despite appearances, Dimetrodon is not a dinosaur.

275 to 201 Million Years Ago:

i. Some pelycosaurs evolve to become the first therapsids. Eventually, the therapsids replace the rest of the pelycosaurs.
ii. About 260 million years ago, a group of therapsids called the cynodonts develops dog-like teeth and eventually evolves into the first mammals.

250 Million Years Ago:

i. The Permian period ends with the greatest mass extinction in Earth's history, wiping out great swathes of species, including the last of the trilobites.
ii. As ecosystems recover, fundamental shifts occur. Whereas before the synapsids (first the pelycosaurs, then the therapsids) dominated the land, the sauropsids now take over – most famously, in the form of dinosaurs. The therapsids survive as small, nocturnal creatures.
iii. In the oceans, the ammonites, cousins of the modern nautilus

and octopus, evolve around this time. Several groups of reptiles colonize the seas, developing into the great marine reptiles of the dinosaur era. The ichthyosaurs adapt rapidly to the marine lifestyle, some becoming as large as modern whales.

233 Million Years Ago:

i. Proto-mammals evolve warm-bloodedness – the ability to maintain their internal temperature, regardless of the external conditions. The timing of this remains controversial – as is whether some or all dinosaurs were also warm-blooded.

210 Million Years Ago:

i. Bird-like footprints and a badly preserved fossil called Protoavis suggest that some early dinosaurs are already evolving into birds at this time. This claim remains controversial.

200 Million Years Ago:

i. As the Triassic period comes to an end, another mass extinction strikes.
ii. In the aftermath, the dinosaurs emerge from among the sauropsids and begin to dominate ecosystems. The largest ichthyosaurs are among the marine casualties of the extinction.

180 Million Years Ago:

i. The first split occurs in the early mammal population. The monotremes, a group of mammals that lay eggs rather than giving birth to live young, break apart from the others. Few monotremes survive today: those that do include the duck-billed platypus and the echidnas.

168 Million Years Ago:

i. A half-feathered, flightless dinosaur called Epidexipteryx, which may be an early step on the road to birds, lives in China.

160 Million Years Ago:

i. Possibly around this time, placental mammals split from their cousins the marsupials. The latter, like the modern kangaroo, give birth when their young are still very small, but nourish them in a pouch for the first few weeks or months of their lives.

ii. The majority of modern marsupials live in Australia, but they may have reached it by an extremely roundabout route. Arising in south-east Asia, they spread into North America, which was attached to Asia at the time. From there they traveled to South America and Antarctica, before making the final journey to Australia about 50 million years ago.

Oldest Evidence Of Pollinating Insects:

i. Some dinosaurs take to the sky with membrane wings like those of modern bats but are soon outcompeted by feathered birds.

150 Million Years Ago:

i. Archaeopteryx, the famous "first bird," lives in Europe.

131 Million Years Ago:

i. Eoconfuciusornis, a bird rather more advanced than Archaeopteryx, lives in China.

125 Million Years Ago:

i. Flowering plants begin to leave their mark in the fossil record, following a period of rapid evolution – although genetics suggests the group is actually older and evolved slowly. The plants go on to dominate the planet, outcompeting the flowerless gymnosperms.

105 to 85 Million Years Ago:

i. The placental mammals split into their four major groups: the laurasiatheres (a hugely diverse group including all the hoofed mammals, whales, bats, and dogs), euarchontoglires (primates, rodents, and others), Xenarthra (including anteaters and armadillos) and afrotheres (elephants, aardvarks, and others).

Quite when, where, and how these splits occurred is unclear at present.

100 Million Years Ago:

i. The Cretaceous dinosaurs reach their peak in size. The giant sauropods Argentinosaurus and Patagotitan, live around this time. They may be the largest land animals in Earth's history – unless an enigmatic sauropod that lived 50 million years earlier was bigger.

93 Million Years Ago:

i. The oceans become starved of oxygen, possibly due to a huge underwater volcanic eruption. Around 27% of marine invertebrates are wiped out.

75 Million Years Ago:

i. The ancestors of modern primates split from the ancestors of modern rodents and lagomorphs (rabbits, hares, and pikas). The rodents go on to be astonishingly successful, eventually making up around 40% of modern mammal species.

70 Million Years Ago:

i. Grasses evolve – though it will be several million years before the vast open grasslands appear.

66 Million Years Ago:

i. The Cretaceous-Tertiary (K/T) extinction wipes out a swathe of species. The giant dinosaurs vanish, although some small forms – the birds – survive. Pterosaurs and plesiosaurs are also among the large reptiles that vanish. The ammonites are also killed off. The extinction clears the way for the mammals, which go on to dominate the planet.

63 Million Years Ago:

i. The primates split into two groups, known as the haplorrhines (dry-nosed primates) and the strepsirrhines (wet-nosed primates). The strepsirrhines eventually become the modern lemurs and aye-ayes, while the haplorrhines develop into monkeys and apes – and humans.

58 Million Years Ago:

i. The tarsiers, primates with enormous eyes to help them see at night, split from the rest of the haplorrhines: the first to do so.

55 Million Years Ago:

i. The Palaeocene-Eocene Thermal Maximum. A sudden rise in greenhouse gases sends temperatures soaring, wiping out many species in the depths of the sea – though sparing species in shallow seas and on land. Archicebus, one of the earliest primates to appear in the fossil record, lives at this time.

50 Million Years Ago:

i. Within the artiodactyls – hoofed mammals including deer and bison – one group begins evolving into whales. Early forms include Pakicetus.

48 Million Years Ago:

i. Indohyus, another artiodactyl closely related to the ancestor of whales and dolphins, lives in India.

47 Million Years Ago:

i. The famous fossilized primate known as "Ida" lives in northern Europe.

ii. Early whales called protocetids live in shallow seas, returning to land to give birth.

40 Million Years Ago:

i. The platyrrhines (sometimes known as the "New World monkeys") become the first simians (higher primates) to diverge from the rest of the group, when they cross the Atlantic and colonize South America.

30 Million Years Ago:

i. Some plants evolve C4 photosynthesis, which is more efficient than other forms. Today major crops like maize are C4 plants.

25 Million Years Ago:

i. Apes split from the monkeys and diversify.

17 Million Years Ago:

i. Gibbons become the first ape to split from the others.

15 Million Years Ago:

i. The ponginae – great apes including the orangutans – branch off from the other great apes, spreading across southern Asia while their cousins remain in Africa. The ponginae include the largest great ape ever to have lived: Gigantopithecus.

7 Million Years Ago:

i. Gorillas branch off from the other great apes.
ii. The ancestors of humans diverge from those of their closest relatives: the chimpanzees and bonobos. Later, some hominins begin walking on two legs.

2.8 Million Years Ago:

i. Our human genus, Homo, appears in Africa.

2 Million Years Ago:

i. A 700-kilogram rodent called *Josephoartigasia monesi* lives in South America. It is the largest rodent known to have lived, displacing the previous record holder: a giant guinea pig.

11,000 Years Ago To Present:

i. Beginning of the Anthropocene, the geological epoch defined by human domination of the planet. The Anthropocene is not an official geological epoch, and the start date remains contentious.

Earlier dates coincide with the beginning of farming and the rise of city living and empires. Alternatively, the start date could be pinned to the 20th century – for example, to the onset of nuclear testing.

ii. The Anthropocene has been marked by a high rate of extinctions, especially of large animals or "megafauna." It may ultimately prove to be the sixth mass extinction. It is also a time of rapid climatic warming, due to greenhouse gas emissions from the burning of fossil fuels and other activities. Like other dramatic shifts before it, the Anthropocene will reshape life on Earth.

Appendix 2: Timeline of Evolutionary Theory

Source: extendedevolutionarysynthesis.com.

1201–1274

Nasīr al-Dīn al-Tūsī

Nasīr al-Dīn al-Tūsī (1201–1274) develops a theory of evolution with organisms gaining differences through adapting to their environments. He suggests that organisms which gain beneficial new features quicker have advantages over others and are more variable.

1735

Carolus Linnaeus

Swedish botanist, Carolus Linnaeus (1707–1778), develops the modern hierarchical classification system.

1749

Georges Buffon

The French naturalist Georges Buffon (1707–1788) envisages a constantly changing world in which species change over time (but rejects the idea that this change could lead to new species).

1773

James Burnett, Lord Monboddo

James Burnett, Lord Monboddo (1767–1792) suggests that humans descended from primates and that creatures can transform their characteristics in response to the environment over long time intervals.

1794

Erasmus Darwin

Erasmus Darwin (1731–1802) proposes that all warm-blooded animals arose and differentiated from a single form and anticipates the idea of natural selection.

1795

Georges Cuvier

Georges Cuvier (1769–1832) draws attention to the fact that the geological record is not a continuous one. He demonstrates the fact of extinction with studies of fossil mammals and believes the extinctions to have occurred in a series of giant floods.

1809

Jean-Baptiste Lamarck

Jean-Baptiste Lamarck (1744–1829) proposes that while simple forms of life were spontaneously generated, they were driven up a ladder of complexity over time. Use or disuse of organs and traits cause changes which could be passed on to the next generation.

1818

Robert Edmond Grant and Étienne Geoffroy Saint-Hilaire

Robert Edmond Grant (1793–1874) and Étienne Geoffroy Saint-Hilaire (1772–1844) further develop the ideas of Jean-Baptiste Lamarck & Erasmus Darwin to propose that plants and animals had a common evolutionary starting point (a view wrongly discredited by Richard Owen).

1830

Charles Lyell

Charles Lyell (1797–1875) establishes the basic chronology of the Tertiary period and its relationship to rock strata. He popularizes the doctrine of uniformitarianism; that the features of the Earth can be better explained as the long-term result of short-term geological phenomena.

1831

Charles Darwin
Charles Darwin (1809–1882) sets off on HMS Beagle.

1845

Alexander von Humboldt
Alexander von Humboldt (1769–1859) pioneers the study of ecology and initiates a new focus on the interactions between species and their environment.

1856

Neanderthal discovery
Neanderthal skull and bones are found in Neander valley in Germany.

1858

Alfred Russel Wallace
Alfred Russel Wallace (1823–1913) independently conceives the theory of evolution by natural selection and co-publishes with Darwin on the subject.

1859

Charles Darwin
Charles Darwin (1809–1882) publishes *On the Origin of Species*.

1860

Oxford Evolution Debate
Proponents and opponents of Darwin & Wallace's theories clash in the famous Oxford Evolution Debate; Thomas Henry Huxley (1825–1895) and Samuel Wilberforce (1805–1873) butt heads in a public debate, which both sides consider a victory.

1866

Gregor Mendel

Gregor Mendel (1822–1884) publishes *Experiments in Plant Hybridization*, establishing some basic laws of the genetic inheritance of discrete traits. Mendel's paper remained obscure for about 35 years, but in 1900 it was rediscovered.

1868

Ernst Haeckel

Ernst Haeckel (1834–1919) applies evolutionary theories to embryology. Subsequently, his work would provide early foundations for the field of evolutionary developmental biology or 'evo devo.'

1871

Charles Darwin

Charles Darwin (1809–1882) publishes *The Descent of Man, and Selection in Relation to Sex*.

1893

August Weissman

August Weissman (1834–1914) publishes his germ-plasm theory, which emphasizes the separation of the germ line and soma.

1896

James Mark Baldwin

James Mark Baldwin (1861–1934) suggests that adaptation can arise and evolve from plasticity without invoking inheritance of acquired characters in *A New Factor in Evolution*, a concept later known as the Baldwin Effect. Similar ideas were also presented earlier or around the same time by Douglas Spalding, Conwy LLoyd Morgan, and Henry Osborn.

1909

Wilhelm Johannsen

Wilhelm Johannsen (1857–1927), a Danish biologist, provides the basic terminology for genetics: 'genes' as particulate units of heredity; 'genotype' as the genetic constitution of an organism; and 'phenotype' as the organism's physical characteristics.

1910

Thomas Hunt Morgan

Thomas Hunt Morgan (1866–1945) establishes the chromosomal theory of heredity. Morgan began to breed and conduct experiments with the common fruit fly, Drosophila melanogaster, which initiated the field of experimental genetics. Morgan confirmed Mendelian laws of inheritance and the hypothesis that genes are located on chromosomes.

1918

Ronald Fisher

Ronald Fisher (1890–1962) produces the paper *The Correlation between Relatives on the Supposition of Mendelian Inheritance*, showing how the continuous variation measured by biometricians could be the result of the action of many discrete genetic loci.

1925

Raymond Dart

Raymond Dart (1893–1988) publishes in Nature a description of the Taung child, a fossil skull from Taung near Johannesburg, of the species *Australopithecus africanus*.

1937

Theodosius Dobzhansky

Theodosius Dobzhansky (1900–1975) publishes *Genetics and the Origin of Species*.

1942

Conrad Hal Waddington

Conrad Hal Waddington (1905–1975) proposes the evolutionary process of genetic assimilation in *Canalization of Development and the Inheritance of Acquired Characters*, followed by several other important papers and books, including *The Strategy of Genes*. His famous epigenetic landscape as a metaphor for gene regulation has become reprinted and re-interpreted ever since.

1942

Julian Huxley
Julian Huxley (1887–1975) publishes *Evolution: The Modern Synthesis*.

1942

Ernst Mayr
Ernst Mayr (1904–2005) publishes *Systematics and the Origin of Species* in which he presents his influential 'biological species concept.'

1950

Radiometric dating developed
Radiometric dating techniques are developed during this decade.

1953

DNA structure discovered
Francis Crick (1916–2004), James Watson (1928-) and Rosalind Franklin (1920–1958) discover the chemical structure of deoxyribonucleic acid (DNA) and show that it meets the unique requirements for a substance that encodes genetic information.

1960

mRNA discovered
Sydney Brenner (1927-), Francis Crick (1916–2004), Francois Jacob (1920–2013) and Jacques Monod (1910–1976) discover mRNA.

1964

William Hamilton
William Hamilton (1936–2000) publishes his papers introducing inclusive fitness theory.

1974

Discovery of 'Lucy'
Paleontologists Don Johanson (1943-) and colleagues find 'Lucy,' an almost complete Australopithecine female at Hadar, Ethiopia.

1975

Allan Wilson and Marie-Claire King
Berkeley biochemists Allan Wilson (1934–1991) and Marie-Claire King (1946-) show that humans share nearly 99% of their DNA with chimpanzees.

1984

Charles Sibley and Jon Ahlquist
Charles Sibley (1917–1998) and Jon Ahlquist (1944-) demonstrate that chimpanzees are more closely related to humans than to gorillas.

1990–2003

Human genome sequenced
The human genome is sequenced and assembled.

2003

Mary Jane West-Eberhard
Mary Jane West-Eberhard (1941-) publishes *Developmental Plasticity and Evolution*, which focuses on the role of environmentally generated variation in evolution and speciation.

2010

Neanderthal genome sequenced
Richard Green and colleagues publish a draft sequence of the Neanderthal genome, suggesting that that Neanderthals and modern humans interbred.

2018

The Present
Exciting evolutionary biology research continues to this day.

Appendix 3: Stages of Evolution of Human Beings

The evolution of man began about 15 million years ago when the first known man walked this earth. Humans today developed through many stages of evolution from primates that are now extinct. This evolutionary process from the primates who walked on all four limbs to the humans today who walk on two hind limbs has been a very long one.

The genus of the human being today is called Homo and the man today is called as Homo sapiens. From simple life forms that were unicellular to the development of multicellular organisms gave rise to the vertebrates. The vertebrates began evolving that led to the development of mammals. Among the mammals, humans are most closely related to primates such as the orangutan.

The family to which human beings belong is called Hominidae. It was in the Miocene age that the family Hominidae split from the Pongidae(apes) family. Dryopethicus was the first in the evolution of man in the stages of evolution and some believe him to be the common ancestor of man and apes.

1. ***Dryopethicus***: He was the earliest known ancestor of man. At the same time as his existence, Ramapethicus existed who was more human-like than Dryopethicus. Dryopethicus inhabited the European region and some parts of Asia and Africa. The stages of evolution of humans began from it. After Dryopethicus and Ramapethicus came to the genus Australopethicus which preceded the genus Homo.

2. ***Australopithecus***

   i. ***Australopithecus ramidus***: Was 1.2 meters tall and the fossils show the foramen magnum that was large to indicate upright walking. The forelimbs were different from those of the earlier ape-like ancestors. They had teeth like humans.

   ii. ***Australopithecus afarensis***: 'Lucy' the famous fossil belonged to this species. They are said to have inhabited the African mainland. They were shorter than the Australopithecus ramidus and had a small skull with flat noses and no chin. They were able to walk on two legs, but the legs were slightly bowed which made their walk slightly ape-like. The bowed legs, fingers, and toes enabled them to climb trees and live there. They had large teeth and jaws.

   iii. ***Australopithecus africanus***: These also inhabited the African mainland. They were bipedal and had a small skull with small brains than Homo erectus but larger than their predecessors. Also, they had large teeth compared to current-day humans and were herbivorous. They had large jaws.

   iv. ***Australopithecus robustus***: He was taller than his predecessors but still ape-like. They also weighed more than their ancestors. After the Australopithecus genus came the Homo genus. The first man in the genus was Homo habilis.

2. Homo:

   i. ***Homo habilis***: He had a face similar to his ancestors. The skull and brain size indicate that he may have been able to speak. The earliest tools made were from this era. Homo habilis is known as the 'handy man' because he was the first to make and use tools. He was around 5 feet tall and erect.

   ii. ***Homo erectus***: After *Homo habilis* came, the Homo erectus who was also upright. He had a smaller but longer face, less prominent or absent chin, larger brain size and prominent speech. He knew how to make and use tools; he made a fire and knew how to control it. Homo erectus was carnivorous. He knew the existence of groups and they began spreading from Africa to Asia and Europe. The Java Man and Peking

Man had brain capacities similar to modern man at 1300cc. They were cave dwellers.

iii. ***Homo sapiens*****:** After Homo erectus came, the Homo sapiens who separated into two types:

a. ***Homo sapiens neanderthelensis*****:** They had a brain size larger than modern man and were gigantic in size. Also, they had a large head and jaw and were very powerful and muscular. They were carnivores and the tools from the era indicate they were hunters. They were also cave dwellers, but their caves were more comfortable, and they lived in groups and hunted for food gathering.

b. ***Homo sapiens sapiens*****:** Also known as 'modern-day man' is what we are today. Compared to the Homo sapiens neanderthelensis, they became smaller in size and the brain size reduced to 1300cc. There was also a reduction in the size of the jaw, rounding of the skull and chin. Cro-Magnon was the earliest of the Homo sapiens. They spread wider from to Europe, Australia, and the Americas. They were omnivores, had skillful hands, developed the power of thinking, produced art, more sophisticated tools, and sentiments.

Evolution is not a thing of the past and is continuing even now. Humans are undergoing 'natural selection' for many different traits based on their life and environment in the present. It is believed that the jaw size is reducing further, and the wisdom teeth are soon going to become extinct.

Glossary

A

Acquired Trait – A phenotypic characteristic, acquired during growth and development, that is not genetically based and therefore cannot be passed on to the next generation (e.g., the large muscles of a weightlifter).

Adaptation – Any heritable characteristic of an organism that improves its ability to survive and reproduce in its environment. Also used to describe the process of genetic change within a population, as influenced by natural selection.

Adaptive Landscape – A graph of the average fitness of a population in relation to the frequencies of genotypes in it. Peaks on the landscape correspond to genotypic frequencies at which the average fitness is high, valleys to genotypic frequencies at which the average fitness is low. Also called a fitness surface.

Adaptive Logic – A behavior has adaptive logic if it tends to increase the number of offspring that an individual contributes to the next and following generations. If such a behavior is even partly genetically determined, it will tend to become widespread in the population. Then, even if circumstances change such that it no longer provides any survival or reproductive advantage, the behavior will still tend to be exhibited – unless it becomes positively disadvantageous in the new environment.

Adaptive Radiation – The diversification, over evolutionary time, of a species or group of species into several different species or subspecies that are typically adapted to different ecological niches (e.g., Darwin's finches).

The term can also be applied to larger groups of organisms, as in "the adaptive radiation of mammals."

Adaptive Strategies – A mode of coping with competition or environmental conditions on an evolutionary time scale. Species adapt when succeeding generations emphasize beneficial characteristics.

Agnostic – A person who believes that the existence of a god or creator and the nature of the universe is unknowable.

Algae – An umbrella term for various simple organisms that contain chlorophyll (and can therefore carry out photosynthesis) and live in aquatic habitats and in moist situations on land. The term has no direct taxonomic significance. Algae range from macroscopic seaweeds such as giant kelp, which frequently exceeds 30 m in length, to microscopic filamentous and single-celled forms such as Spirogyra and Chlorella.

Allele – One of the alternative forms of a gene. For example, if a gene determines the seed color of peas, one allele of that gene may produce green seeds and another allele produce yellow seeds. In a diploid cell, there are usually two alleles of any one gene (one from each parent). Within a population there may be many different alleles of a gene; each has a unique nucleotide sequence.

Allometry – The relation between the size of an organism and the size of any of its parts. For example, an allometric relation exists between brain size and body size, such that (in this case) animals with bigger bodies tend to have bigger brains. Allometric relations can be studied during the growth of a single organism, between different organisms within a species, or between organisms in different species.

Allopatric Speciation – Speciation that occurs when two or more populations of a species are geographically isolated from one another sufficiently that they do not interbreed.

Allopatry – Living in separate places. Compare with sympatry.

Amino Acid – The unit molecular building block of proteins, which are chains of amino acids in a certain sequence. There are 20 main amino acids in the proteins of living things, and the properties of a protein are determined by its particular amino acid sequence.

Amino Acid Sequence – A series of amino acids, the building blocks of proteins, usually coded for by DNA. Exceptions are those coded for by the RNA of certain viruses, such as HIV.

Ammonoid – Extinct relatives of cephalopods (squid, octopi, and chambered nautiluses), these mollusks had coiled shells and are found in the fossil record of the Cretaceous period.

Amniotes – The group of reptiles, birds, and mammals. These all develop through an embryo that is enclosed within a membrane called an amnion. The amnion surrounds the embryo with a watery substance and is probably an adaptation for breeding on land.

Amphibians – The class of vertebrates that contains the frogs, toads, newts, and salamanders. The amphibians evolved in the Devonian period (about 370 million years ago) as the first vertebrates to occupy the land. They have moist scaleless skin which is used to supplement the lungs in gas exchange. The eggs are soft and vulnerable to drying, therefore reproduction commonly occurs in water. Amphibian larvae are aquatic and have gills for respiration; they undergo metamorphosis to the adult form. Most amphibians are found in damp environments, and they occur on all continents except Antarctica.

Analogous Structures – Structures in different species that look alike or perform similar functions (e.g., the wings of butterflies and the wings of birds) that have evolved convergently but do not develop from similar groups of embryological tissues, and that have not evolved from similar structures known to be shared by common ancestors. Contrast with homologous structures. Note: The recent discovery of deep genetic homologies has brought new interest, new information, and discussion to the classical concepts of analogous and homologous structures.

Anatomy – (1) The structure of an organism or one of its parts. (2) The science that studies those structures.

Ancestral Homology – Homology that evolved before the common ancestor of a set of species, and which is present in other species outside that set of species. Compare with derived homology.

Anthropoid – A member of the group of primates made up of monkeys, apes, and humans.

Antibacterial – Having the ability to kill bacteria.

Antibiotic Resistance – A heritable trait in microorganisms that enables them to survive in the presence of an antibiotic.

Antibiotics – Substances that destroy or inhibit the growth of microorganisms, particularly disease-causing bacteria.

Aperture – Of a camera, the adjustable opening through which light passes to reach the film. The diameter of the aperture determines the intensity of light admitted. The pupil of a human eye is a self-adjusting aperture.

Aquatic – Living underwater.

Arboreal – Living in trees.

Archeology – The study of human history and prehistory through the excavation of sites and the analysis of physical remains, such as graves, tools, pottery, and other artifacts.

Archetype – The original form or body plan from which a group of organisms develops.

Artifact – An object made by humans that has been preserved and can be studied to learn about a particular time period.

Artificial Selection – The process by which humans breed animals and cultivate crops to ensure that future generations have specific desirable characteristics. In artificial selection, breeders select the most desirable variants in a plant or animal population and selectively breed them with other desirable individuals. The forms of most domesticated and agricultural species have been produced by artificial selection; it is also an important experimental technique for studying evolution.

Asexual Reproduction – A type of reproduction involving only one parent that usually produces genetically identical offspring. Asexual reproduction occurs without fertilization or genetic recombination, and may occur by budding, by division of a single cell, or by the breakup of a whole organism into two or more new individuals.

Assortative Mating – The tendency of like to mate with like. Mating can be assortative for a certain genotype (e.g., individuals with genotype AA tend to mate with other individuals of genotype AA) or phenotype (e.g., tall individuals' mate with other tall individuals).

Asteroid – A small, rocky, or metallic body orbiting the Sun. About 20,000 have been observed, ranging in size from several 100 kilometers across down to dust particles.

Atheism – The doctrine or belief that there is no god.

Atomistic – (as applied to the theory of inheritance) Inheritance in which the entities controlling heredity are relatively distinct, permanent, and capable of independent action. Mendelian inheritance is an atomistic theory because in it, inheritance is controlled by distinct genes.

Australopithecine – A group of bipedal hominid species belonging to the genus Australopithecus that lived between 4.2 and 1.4 mya.

Australopithecus afarensis – An early australopithecine species that was bipedal; known fossils date between 3.6 and 2.9 mya (e.g., Lucy).

Autosome –Any chromosome other than a sex chromosome.

Avian – Of, relating to, or characteristic of birds (members of the class Aves).

B

Bacteria – Tiny, single-celled, prokaryotic organisms that can survive in a wide variety of environments. Some cause serious infectious diseases in humans, other animals, and plants.

Base – The DNA molecule is a chain of nucleotide units; each unit consists of a backbone made of a sugar and a phosphate group, with a nitrogenous base attached. The base in a unit is one of adenine (A), guanine (G), cytosine (C), or thymine (T). In RNA, uracil (U) is used instead of thymine. A and G belong to the chemical class called purines; C, T, and U are pyrimidines.

Batesian Mimicry – A kind of mimicry in which one non-poisonous species (the Batesian mimic) mimics another poisonous species.

Belemnite – An extinct marine invertebrate that was related to squid, octopi, and chambered nautiluses. We know from the fossil record that belemnites were common in the Jurassic period and had bullet-shaped internal skeletons.

Big Bang theory – The theory that states that the universe began in a state of compression to infinite density, and that in one instant all matter and energy began expanding and have continued expanding ever since.

Biodiversity (or Biological Diversity) – A measure of the variety of life, biodiversity is often described on three levels. Ecosystem diversity describes the variety of habitats present; species diversity is a measure of the number of species and the number of individuals of each species present; genetic diversity refers to the total amount of genetic variability present.

Bioengineered Food – Food that has been produced through genetic modification using techniques of genetic engineering.

Biogenetic Law – Name given by Haeckel to recapitulation.

Biogeography – The study of patterns of the geographical distribution of plants and animals across Earth, and the changes in those distributions over time.

Biological Species Concept – The concept of species, according to which a species is a set of organisms that can interbreed among each other. Compare with cladistic species concept, ecological species concept, phenetic species concept, and recognition species concept.

Biometrics – The quantitative study of characters of organisms.

Biosphere – The part of Earth and its atmosphere capable of sustaining life.

Bipedalism – Of hominids, walking upright on two hind legs; more generally, using two legs for locomotion.

Bivalve – A mollusk that has a two-part hinged shell. Bivalves include clams, oysters, scallops, mussels, and other shellfish.

Blackmore, Susan – A psychologist interested in memes and the theory of memetics, evolutionary theory, consciousness, the effects of meditation, and why people believe in the paranormal. A recent book, The Meme Machine, offers an introduction to the subject of memes.

Blending Inheritance – The historically influential but factually erroneous theory that organisms contain a blend of their parents' hereditary factors and pass that blend on to their offspring. Compare with Mendelian inheritance.

Botanist – A scientist who studies plants.

Brachiopod – Commonly known as "lamp shells," these marine invertebrates resemble bivalve mollusks because of their hinged shells. Brachiopods were at their greatest abundance during the Paleozoic and Mesozoic eras.

Brodie, Edmund D., III – A biologist who studies the causes and evolutionary implications of interactions among traits in predators and their prey. Much of his work concentrates on the coevolutionary arms race between newts that possess tetrodotoxin, one of the most potent known toxins, and the resistant garter snakes that prey on them.

Brodie, Edmund D., Jr. – A biologist recognized internationally for his work on the evolution of mechanisms in amphibians that allow them to avoid predators. These mechanisms include toxins carried in skin secretions, coloration, and behavior.

Bruner, Jerome – A psychologist and professor at Harvard and Oxford Universities, and a prolific author whose book, The Process of Education, encouraged curriculum innovation based on theories of cognitive development.

Bryozoan – A tiny marine invertebrate that forms a crust-like colony; colonies of bryozoans may look like scaly sheets on seaweed.

Burney, David – A biologist whose research has focused on endangered species, paleoenvironmental studies, and causes of extinction in North

America, Africa, Madagascar, Hawaii, and the West Indies.

C

Carbon Isotope Ratio – A measure of the proportion of the carbon-14 isotope to the carbon-12 isotope. Living material contains carbon-14 and carbon-12 in the same proportions as exists in the atmosphere. When an organism dies, however, it no longer takes up carbon from the atmosphere, and the carbon-14 it contains decays to nitrogen-14 at a constant rate. By measuring the carbon-14-to-carbon-12 ratio in a fossil or organic artifact, its age can be determined, a method called radiocarbon dating. Because most carbon-14 will have decayed after 50,000 years, the carbon isotope ratio is mainly useful for dating fossils and artifacts younger than this. It cannot be used to determine the age of Earth, for example.

Carnivorous – Feeding largely or exclusively on meat or other animal tissue.

Carroll, Sean – Developmental geneticist with the Howard Hughes Medical Institute and professor at the University of Wisconsin-Madison. From the large-scale changes that distinguish major animal groups to the finely detailed color patterns on butterfly wings, Dr. Carroll's research has centered on those genes that create the "molecular blueprint" for body pattern and play major roles in the origin of new features. Coauthor, with Jennifer Grenier and Scott Weatherbee, of From DNA to Diversity: Molecular Genetics and the Evolution of Animal Design.

Carson, Rachel – A scientist and writer fascinated with the workings of nature. Her best-known publication, Silent Spring, was written over the years 1958 to 1962. The book looks at the effects of insecticides and pesticides on songbird populations throughout the United States. The publication helped set off a wave of environmental legislation and galvanized the emerging ecological movement.

Castle, W. E. – An early experimental geneticist, his 1901 paper was the first on Mendelism in America. His Genetics of Domestic Rabbits, published in 1930 by Harvard University Press, covers such topics as the genes involved in determining the coat colors of rabbits and associated mutations.

Cell – The basic structural and functional unit of most living organisms. Cell size varies, but most cells are microscopic. Cells may exist as independent units of life, as in bacteria and protozoans, or they may form colonies or tissues, as in all plants and animals. Each cell consists of a mass of protein material that is differentiated into cytoplasm and nucleoplasm, which contains DNA. The cell is enclosed by a cell membrane, which in the cells of plants, fungi, algae, and bacteria is surrounded by a cell wall. There are two main types of cells, prokaryotic, and eukaryotic.

Cenozoic – The era of geologic time from 65 mya to the present, a time when the modern continents formed, and modern animals and plants evolved.

Centromere – A point on a chromosome that is involved in separating the copies of the chromosome produced during cell division. During this division, paired chromosomes look somewhat like an X, and the centromere is the constriction in the center.

Cephalopod – Cephalopods include squid, octopi, cuttlefish, and chambered nautiluses. They are mollusks with tentacles and move by forcing water through their bodies like a jet.

Character – Any recognizable trait, feature, or property of an organism. In phylogenetic studies, a character is a feature that is thought to vary independently of other features, and to be derived from a corresponding feature in a common ancestor of the organisms being studied. A "character state" is one of the possible alternative conditions of the character. For example, "present" and "absent" are two states of the character "hair" in mammals. Similarly, a particular position in a DNA sequence is a character, and A, T, C, and G are its possible states (see bases.)

Character Displacement – The increased difference between two closely related species where they live in the same geographic region (sympatry) as compared with where they live in different geographic regions (allopatry). Explained by the relative influences of intra- and inter-specific competition in sympatry and allopatry.

Chloroplast – A structure (or organelle) found in some cells of plants; its function is photosynthesis.

Cholera – An acute infectious disease of the small intestine, caused by the bacterium *Vibrio cholerae* which is transmitted in drinking water contaminated by the feces of a patient. After an incubation period of 1–5 days, cholera causes severe vomiting and diarrhea, which, if untreated, leads to dehydration that can be fatal.

Chordate – A member of the phylum Chordata, which includes the tunicates, lancelets, and vertebrates. They are animals with a hollow dorsal nerve cord; a rodlike notochord that forms the basis of the internal skeleton; and paired gill slits in the wall of the pharynx behind the head, although in some chordates these are apparent only in early embryonic stages. All vertebrates are chordates, but the phylum also contains simpler types, such as sea-squirts, in which only the free-swimming larva has a notochord.

Chromosomal Inversion – See inversion.

Chromosome – A structure in the cell nucleus that carries DNA. At certain times in the cell cycle, chromosomes are visible as string-like entities. Chromosomes consist of the DNA with various proteins, particularly histones, bound to it.

Chronology – The order of events according to time.

Clack, Jenny – A paleontologist at Cambridge University in the U.K., Dr. Clack studies the origin, phylogeny, and radiation of early tetrapods and their relatives among the lobe-finned fish. She is interested in the timing and sequence of skeletal and other changes that occurred during the transition, and the origin and relationships of the diverse tetrapods of the late Paleozoic.

Clade – A set of species descended from a common ancestral species. Synonym of monophyletic group.

Cladism – Phylogenetic classification. The members of a group in a cladistic classification share a more recent common ancestor with one another than with the members of any other group. A group at any level in the classificatory hierarchy, such as a family, is formed by combining a subgroup at the next lowest level (the genus, in this case) with the subgroup or subgroups with which it shares its most recent common ancestor. Compare with evolutionary classification and phenetic classification.

Cladistic Species Concept – The concept of species, according to which a species is a lineage of populations between two phylogenetic branch points (or speciation events). Compare with biological species concept, ecological species concept, phenetic species concept, and recognition species concept.

Cladists – Evolutionary biologists who seek to classify Earth's life forms according to their evolutionary relationships, not just overall similarity.

Cladogram – A branching diagram that illustrates hypotheses about the evolutionary relationships among groups of organisms. Cladograms can be considered as a special type of phylogenetic tree that concentrates on the order in which different groups branched off from their common ancestors. A cladogram branches like a family tree, with the most closely related species on adjacent branches.

Class – A category of taxonomic classification between order and phylum, a class comprises members of similar orders. See taxon.

Classification – The arrangement of organisms into hierarchical groups. Modern biological classifications are Linnaean and classify organisms into species, genus, family, order, class, phylum, kingdom, and certain intermediate categoric levels. Cladism, evolutionary classification, and phenetic classification are three methods of classification.

Cline – A geographic gradient in the frequency of a gene, or in the average value of a character.

Clock – See molecular clock.

Clone – A set of genetically identical organisms asexually reproduced from one ancestral organism.

Coadaptation – Beneficial interaction between (1) a number of genes at different loci within an organism, (2) different parts of an organism, or (3) organisms belonging to different species.

Codon – A triplet of bases (or nucleotides) in the DNA coding for one amino acid. The relation between codons and amino acids is given by the genetic code. The triplet of bases that is complementary to a Condon is called an anticodon; conventionally, the triplet in the mRNA is called the codon and the triplet in the tRNA is called the anticodon.

Coelacanth – Although long thought to have gone extinct about 65 million years ago, one of these deep-water, lungless fish was caught in the 1930s. Others have since been caught and filmed in their natural habitat.

Coevolution – Evolution in two or more species, such as predator and its prey or a parasite and its host, in which evolutionary changes in one species influence the evolution of the other species.

Cognitive – Relating to cognition, the mental processes involved in the gathering, organization, and use of knowledge, including such aspects as awareness, perception, reasoning, and judgement. The term refers to any mental "behaviors" where the underlying characteristics are abstract in nature and involve insight, expectancy, complex rule use, imagery, use of symbols, belief, intentionality, problem-solving, and so forth.

Common Ancestor – The most recent ancestral form or species from which two different species evolved.

Comparative Biology – The study of patterns among more than one species.

Comparative Method – The study of adaptation by comparing many species.

Concerted Evolution – The tendency of the different genes in a gene family to evolve in concert; that is, each gene locus in the family comes to have the same genetic variant.

Conodont – A jawless fish that had tiny, tooth-like phosphate pieces that are abundant in the fossil record, these were the earliest known vertebrates.

Continental Drift – The process by which the continents move as part of large plates floating on Earth's mantle. See plate tectonics.

Contrivance – An object or characteristic used or modified to do something different from its usual use.

Convergence – The process by which a similar character evolves independently in two species. Also, a synonym for analogy; that is, an instance of a convergently evolved character, or a similar character in two species that was not present in their common ancestor. Examples include wings (convergent in birds, bats, and insects) and camera-type eyes (convergent in vertebrates and cephalopod mollusks).

Convergent Evolution – The evolution of species from different taxonomic groups toward a similar form; the development of similar characteristics by taxonomically different organisms.

Conway Morris, Simon – Paleobiologist and professor in the Department of Earth Sciences at Cambridge University in the U.K. His research centers around the early evolution of the metazoans, and he is a leading authority on Cambrian and Precambrian fossils. Conway Morris established a link between the Ediacaran fossils, a Burgess Shale fernlike frond Thaumaptilon, and the modern seapens, colonial animals related to the corals.

Cope's Rule – The evolutionary increase in body size over geological time in a lineage of populations.

Coral (also, Rugose Coral, Tabulate Coral) – These tiny animals make calcium carbonate skeletons that are well-known as a key part of tropical reefs. The skeletons of the extinct rugose and tabulate corals are known from fossils.

Cranium – The part of the skull that protects the brain in vertebrates.

Creation Science – An assortment of many different, non-scientific attempts to disprove evolutionary theory, and efforts to prove that the complexity of living things can be explained only by the action of an "intelligent designer."

Creationism – The religious doctrine that all living things on Earth were created separately, in more or less their present form, by a supernatural creator, as stated in the Bible; the precise beliefs of different creationist groups vary widely. See separate creation.

Cretaceous – The final geological period of the Mesozoic era that began 144 million years ago and ended 65 million years ago. The end of this period is defined most notably by the extinction of the dinosaurs in one of the largest mass extinctions ever to strike the planet.

Crinoid – A marine invertebrate animal belonging to a class (Crinoidea; about 700 species) of echinoderms, including sea lilies and feather stars. They have a small cup-shaped body covered with hard plates and five radiating pairs of feathery flexible arms surrounding the mouth at the top. Sea lilies, most of which are extinct, are fixed to the sea bottom or some other surface such as a reef by a stalk. Feather stars are free-swimming and are usually found on rocky bottoms. Crinoids occur mainly in deep waters and feed on microscopic plankton and detritus caught by the arms and conveyed to the mouth. The larvae are sedentary. They arose in the Lower Ordovician (between 500 and 460 million years ago), and fossil crinoids are an important constituent of Paleozoic limestones.

Crossing Over – The process during meiosis in which the chromosome of a diploid pair exchange genetic material, visible in the light microscope. At a genetic level, it produces recombination.

Crustacean – A group of marine invertebrates with exoskeletons and several pairs of legs. They include shrimp, lobsters, crabs, amphipods (commonly known as "sand fleas"), and many more.

Currie, Cameron – A Canadian ecologist and recipient of the 2001 Natural Sciences and Engineering Research Council Doctoral Prize for his research on the complex symbiotic relationship of fungus-growing ants, the fungi they cultivate, mutualistic bacteria that the ants carry on their bodies, and pathogens that attack the fungi.

Cytoplasm – The region of a eukaryotic cell outside the nucleus.

D

Daeschler, Ted – Paleontologist and associate research curator at the Academy of Natural Sciences. Discoverer of late Devonian limbed fossils *Hynerpeton bassetti* and *Designathusro wei* (tetrapods) and *Sauripterus taylori* and Hyneria (lobed-finned fishes), all early examples of animals exploiting both land and water environments. Author of two books on paleontology for young people.

Dart, Raymond – Australian-born South African anatomist and anthropologist (1893–1988). In 1924 he described a fossil skull collected near Taung in South Africa, naming it Australopithecus africanus. Dart asserted that the skull was intermediate between the apes and humans, a controversial claim at the time, though later work made it clear that the Taung child, as it came to be known, was indeed a hominid.

Darwin, Charles – The 19th-century naturalist considered the father of evolution. His landmark work, On the Origin of Species, published in 1859, presented a wealth of facts supporting the idea of evolution and proposed a viable theory for how evolution occurs – via the mechanism Darwin called "natural selection." In addition to his prolific work in biology, Darwin also published important works on coral reefs and on the geology of the Andes, and a popular travelogue of his five-year voyage aboard HMS Beagle.

Darwin, Erasmus – The name shared by Charles Darwin's grandfather and brother, each important in his life and work. Charles's grandfather Erasmus (1731–1802) was a glorious polymath – physician, author, and botanist. His impact is reflected throughout a wide range of disciplines from the poetry to the technology of his day. Author of The Loves of the Plants, a 2,000-line poem detailing their sexual reproduction, and Zoonomia, or the Theory of Generations, whose themes echo throughout his grandson's work. Charles's older brother Erasmus (1805–1881), known as "Ras," used his network of social and scientific contacts to advance the theories of his shyer, more retiring sibling.

Darwinian Evolution – Evolution by the process of natural selection acting on random variation.

Darwinism – Darwin's theory that species originated by evolution from other species and that evolution is mainly driven by natural selection. Differs from Neo-Darwinism mainly in that Darwin did not know about Mendelian inheritance.

Dawkins, Richard – An evolutionary biologist who has taught zoology and is the author of several books on evolution and science, including The Selfish Gene (1976) and The Blind Watchmaker (1986). He is known for his popularization of Darwinian ideas, as well as for original thinking on evolutionary theory.

De Facto – In fact; in reality. Something which exists or occurs de facto is not the result of a law, but because of circumstances.

Dembski, William – A mathematician and philosopher who has written on intelligent design, attempting to establish the legitimacy and fruitfulness of design within biology.

Dennett, Daniel – Philosopher and director of the Center for Cognitive Studies at Tufts University, whose work unites neuroscience, computer science, and evolutionary biology. Dennett sees no basic distinction between human and machine intelligence, advocating a mechanical explanation of consciousness. He is the author of Brainchildren: Essays on Designing Minds and Darwin's Dangerous Idea: Evolution and the Meanings of Life, among many other books and publications.

Derived Homology – Homology that first evolved in the common ancestor of a set of species and is unique to those species. Compare with ancestral homology.

Diatom – These single-celled algae are common among the marine phytoplankton. Their glassy, two-part shells have intricate patterns and fit together like the two parts of a shirt box.

Diffusion – The process by which molecules (e.g., of oxygen) move passively from a region of high concentration to a region of low concentration.

Dinoflagellate – Possessing two tail-like extensions called flagella that are used for movement, these single-celled algae can live freely or in other organisms such as corals. When many dinoflagellates suddenly reproduce in great numbers, they create what are known as "red tides" by making the water appear red.

Diploid – Having two sets of genes and two sets of chromosomes (one from the mother, one from the father). Many common species, including humans, are diploid. Compare with haploid and polyploid.

Directional Selection – Selection causing a consistent directional change in the form of a population through time (e.g., selection for larger body size).

Disruptive Selection – Selection favoring forms that deviate in either direction from the population average. Selection favors forms that are larger or smaller than average but works against the average forms between the extremes.

Distance – In taxonomy, it refers to the quantitatively measured difference between the phenetic appearance of two groups of individuals, such as populations or species (phenetic distance), or the difference in their gene frequencies (genetic distance).

DNA – Deoxyribonucleic acid, the molecule that controls inheritance.

DNA Base Sequence – A chain of repeating units of deoxyribonucleotides (adenine, guanine, cytosice, thymine) arranged in a particular pattern.

Dobzhansky, Theodosius – A geneticist and zoologist best known for his research in population genetics using the fruit fly. His study of the evolution of races led to the discovery of genetic diversity within species, and confirmed his belief that genetic variation leads to better adaptability.

Dominance (Genetic) – An allele (A) is dominant if the phenotype of the heterozygote (Aa) is the same as the homozygote (AA). The allele (a) does not influence the heterozygote's phenotype and is called recessive. An allele may be partly, rather than fully, dominant; in that case, the heterozygous phenotype is nearer to, rather than identical with, the homozygote of the dominant allele.

Drift – Synonym of genetic drift.

Duplication – The occurrence of a second copy of a particular sequence of DNA. The duplicate sequence may appear next to the original or be copied elsewhere into the genome. When the duplicated sequence is a gene, the event is called gene duplication.

E

Echinoderm – Echinoderms, whose name means "spiny skin," are a group of marine invertebrates that includes starfish, brittle stars, basket stars, sea cucumbers, sand dollars, sea urchins, and others. They live in environments from shallow coastal waters to deep-sea trenches, from the tropics to the poles.

Ecological Genetics – The study of evolution in action in nature, by a combination of field work and laboratory genetics.

Ecological Species Concept – A concept of species, according to which a species is a set of organisms adapted to a particular, discrete set of resources (or "niche") in the environment. Compare with biological species concept, cladistic species concept, phenetic species concept, and recognition species concept.

Ecosystem – A community of organisms interacting with a particular environment.

Eldredge, Niles – A paleontologist and evolutionary biologist with the American Museum of Natural History, Eldredge, together with Stephen Jay Gould, proposed the theory of punctuated equilibria, providing paleontologists with an explanation for the patterns which they find in the fossil record. He has written several books for a general audience, including Time Frames: The Evolution of Punctuated Equilibria and Life in the Balance: Humanity and the Biodiversity Crisis.

Electrophoresis – The method of distinguishing entities according to their motility in an electric field. In evolutionary biology, it has been mainly used to distinguish different forms of proteins. The electrophoretic motility of a molecule is influenced by its size and electric charge.

Embryo – An early stage of animal development that begins after the division of the zygote (the earliest stage, in which joined egg and sperm have not yet divided).

Embryology – scientific study of the earliest stages of growth and development in organisms.

Embryonic – Related to an embryo or being in the state of an embryo.

Emigration – The movement of organisms out of an area.

Emlen, Stephen – A world authority on the social behavior of animals, particularly birds. Emlen's interests center on evolutionary or adaptive aspects of animal behavior. The goal of his research is to better understand the social interactions that occur between individuals, especially cooperation and conflict.

Empirical – Determined by experimentation.

Endler, John – A zoologist and professor with interests in evolution and how it affects geographic variation. His current research focuses on guppies (*Poecilia reticulata*) in their natural habitat, and how visual signs and vision dictate their behavior.

Enzyme – A protein that acts as a catalyst for chemical reactions.

Eocene – The second oldest of the five major epochs of the Tertiary period, from 54 to 38 mya. It is often known for the rise of mammals.

Epistasis – An interaction between the genes at two or more loci, such that the phenotype differs from at would be expected if the loci were expressed independently.

Erwin, Douglas – Dr. Erwin is a paleobiologist with the National Museum of Natural History in the Smithsonian Institution. His research is concerned with aspects of major evolutionary novelties, particularly the Metazoan radiation and post-mass extinction recoveries. Recent work has involved the developmental events associated with the Cambrian along with their environmental context. He also works on the rate, causes, and consequences of the end-Permian mass extinction.

Eugenics – The science or practice of altering a population, especially of humans, by controlled breeding for desirable inherited characteristics. The term was coined in 1883 by Francis Galton, who was an advocate of "improving" the human race by modifying the fertility of different categories of people. Eugenics fell into disfavor after the perversion of its doctrines by the Nazis.

Eukaryote – Any organism made up of eukaryotic cells. Eukaryotes are generally larger and have more DNA than prokaryotes (whose cells do not have a nucleus to contain their DNA). Almost all multicellular organisms are eukaryotes.

Eukaryotic Cell – A cell with a distinct nucleus.

Evo-Devo – The informal name for 'evolutionary developmental biology,' a branch of evolutionary studies that seeks to contribute insights from genetics, embryology, and microbiology to help scientists better understand how new species arise from existing life forms.

Evolution – Darwin defined this term as "descent with modification." It is the change in a lineage of populations between generations. In general terms, biological evolution is the process of change by which new species develop from preexisting species over time; in genetic terms, evolution can be defined as any change in the frequency of alleles in populations of organisms from generation to generation.

Evolutionary Classification – Method of classification using both cladistic and phenetic classificatory principles. To be exact, it permits paraphyletic groups (which are allowed in phenetic but not in cladistic classification) and monophyletic groups (which are allowed in both cladistic and phenetic classification) but excludes polyphyletic groups (which are banned from cladistic classification but permitted in phenetic classification).

Ewald, Paul – Professor of biology at Amherst College, specializing in hummingbird and flower coevolution and the evolution of infectious diseases. His research on disease focuses on the evolutionary effects of various public health interventions.

Exon – The nucleotide sequences of some genes consist of parts that code for amino acids, with other parts that do not code for amino acids interspersed among them. The coding parts, which are translated, are called exons; the interspersed non-coding parts are called introns.

Extinction – The disappearance of a species or a population.

F

Fact – A natural phenomenon repeatedly confirmed by observation.

Family – The category of taxonomic classification between order and genus (see taxon). Organisms within a family share a close similarity; for example, the cat family, Felidae, which includes lions and domestic cats.

Fauna – Animal life; often used to distinguish from plant life ("flora").

Fermentation – A series of reactions occurring under anaerobic conditions (lacking oxygen) in certain microorganisms (particularly yeasts) in which organic compounds such as glucose are converted into simpler substances with the release of energy. Fermentation is involved in bread making, where the carbon dioxide produced by the yeast causes dough to rise.

Fetus – The embryo of a mammal that has reached a stage of development in the uterus in which most of the adult features are recognizable. Specifically in humans, it refers to the stage of development after the appearance of bone cells, a process occurring 7 to 8 weeks after fertilization.

Fitness – The success of an individual (or allele or genotype in a population) in surviving and reproducing, measured by that individual's (or alleles or genotype's) genetic contribution to the next generation and subsequent generations.

FitzRoy, Robert – Captain of the Beagle, which took Charles Darwin on his famous voyage to South America and around the world. FitzRoy's chief mission on the Beagle was to chart the coast of South America. He also established the first weather warning system while on his journeys, with the help of the telegraph, and later rose to the rank of Admiral in the British Navy. He was known as a young man for his moody temperament, and in his older age for questionable sanity; FitzRoy's life ended in suicide.

Fixation – A gene has achieved fixation when its frequency has reached 100% in the population.

Fixed – (1) In population genetics, a gene is "fixed" when it has a frequency of 100%. (2) In creationism, species are described as "fixed" in the sense that they are believed not to change their form, or appearance, through time.

Flammer, Larry – A retired high school biology teacher and co-founder of the Santa Clara County Biotechnology-Education Partnership, which provides teacher training and lab equipment for local schools. He is a current member and Web writer for the Evolution and Nature of Science Institute (ENSI).

Flora – Plant life; often used to distinguish from animal life ("fauna").

Foraminifera – These invertebrates are very common in the global ocean, and their distinctive, chambered shells are common in the fossil record as far back as 550 million years. Although very few today exceed 9 mm in diameter, fossils have been found that measure 15 cm across.

Fossil – Most commonly, an organism, a physical part of an organism, or an imprint of an organism that has been preserved from ancient times in rock, amber, or by some other means. New techniques have also revealed the existence of cellular and molecular fossils.

Founder Effect – The loss of genetic variation when a new colony is formed by a very small number of individuals from a larger population.

Frequency-Dependent Selection – Selection in which the fitness of a genotype (or phenotype) depends on its frequency in the population.

Fungi – A group of organisms comprising the kingdom Fungi, which includes molds and mushrooms. They can exist either as single cells or make up a multicellular body called a mycelium. Fungi lack chlorophyll and secrete digestive enzymes that decompose other biological tissues.

G

Galton, Francis – A cousin of Charles Darwin, Galton was a British explorer and anthropologist. He was known for his studies of human intelligence and later for his work in eugenics (a term he coined), the "science" of improving human hereditary characteristics. Known for his efforts at various sorts of measurement (he developed fingerprinting and was a pioneer in statistics), he was knighted in 1909.

Gamete – Haploid reproductive cells that combine at fertilization to form the zygote, called sperm (or pollen) in males and eggs in females.

Gastropod – Meaning "stomach foot," this name refers to the class of mollusks that contains the most species. Gastropods include snails and slugs that are marine, freshwater, and terrestrial.

Gehring, Walter J. – Dr. Gehring and his research group discovered the homeobox, a DNA segment characteristic for homeotic genes that is not only present in arthropods and their ancestors, but also in vertebrates up to humans. Their work on the "master control gene" for eye development sheds light on how the mechanism for building eyes may have evolved long ago in the ancestor of what are now very different types of organisms.

Gene – A sequence of nucleotides coding for a protein (or, in some cases, part of a protein); a unit of heredity.

Gene Duplication – See duplication.

Gene Family – A set of related genes occupying various loci in the DNA, almost certainly formed by duplication of an ancestral gene, and having a recognizably similar sequence. Members of a gene family may be functionally very similar or differ widely. The globin gene family is an example.

Gene Flow – The movement of genes into or through a population by interbreeding or by migration and interbreeding.

Gene Frequency – The frequency in the population of a particular gene relative to other genes at its locus. Expressed as a proportion (between 0 and 1) or percentage (between 0 and 100%).

Gene Pool – All the genes in a population at a particular time.

Genetic – Related to genes. A gene is a sequence of nucleotides coding for a protein (or, in some cases, part of a protein), a unit of heredity.

Genetic Code – The code relating nucleotide triplets in the mRNA (or DNA) to amino acids in the proteins.

Genetic Distance – See distance.

Genetic Drift – Changes in the frequencies of alleles in a population that occur by chance, rather than because of natural selection.

Genetic Engineering – Removing genes from the DNA of one species and splicing them into the DNA of another species using the techniques of molecular biology.

Genetic Load – A reduction in the average fitness of the members of a population because of the deleterious genes, or gene combinations, in the population. It has many particular forms, such as "mutational load," "segregational load," and "recombinational load."

Genetic Locus – See locus.

Genetics – The study of genes and their relationship to characteristics of organisms.

Genome – The full set of DNA in a cell or organism.

Genomics – The study that characterizes genes and the traits they encode.

Genotype – The set of two genes possessed by an individual at a given locus. More generally, the genetic profile of an individual.

Genus (Plural Genera) – The second-to-lowest category in taxonomic classification. The phrase "species name" generally refers to the genus and species together, as in the Latin name for humans, Homo sapiens. See taxon.

Geographic Isolation – See reproductive isolation.

Geographic Speciation – See allopatric speciation.

Geologic Time – The time scale used to describe events in the history of Earth.

Germination – The initial stages in the growth of a seed to form a seedling. The embryonic shoot (plumule) and embryonic root (radicle) emerge and grow upward and downward, respectively. Food reserves for germination come from tissue within the seed and/or from the seed leaves (cotyledons).

Germplasm – The reproductive cells in an organism, or the cells that produce the gametes. All cells in an organism can be divided into the soma (the cells that ultimately die) and the germ cells (the cells that are perpetuated by reproduction).

Gestation – The period in animals bearing live young (especially mammals) from the fertilization of the egg and its implantation into the wall of the uterus until the birth of the young (parturition), during which the young develops in the uterus. In humans, gestation is known as pregnancy and takes about nine months (40 weeks).

Gingerich, Philip – Gingerich is interested in evolutionary change documented in the fossil record and how this relates to the kinds of changes observable in the field or laboratory on the scale of a few generations. His ongoing fieldwork in Wyoming, Egypt, and Pakistan is concerned with the origin of modern orders of mammals, especially primates and whales.

Glaciation – The formation of large sheets of ice across land. Glaciation of the continents marks the beginning of ice ages when the makeup of Earth and organisms on it changes dramatically.

Goldfarb, Alex – A Russian-born microbiologist now at the Public Health Research Institute in New York City, Dr. Goldfarb is piloting a program in the Russian prison system to combat the further evolution of drug-resistant strains of tuberculosis, which have infected at least 30,000 inmates.

Gould, Stephen Jay – A professor of geology and zoology at Harvard University since 1967. A paleontologist and an evolutionary biologist, he teaches geology and the history of science, as well. With others, he has advanced the concept that major evolutionary changes can occur in sudden bursts rather than through the slow, gradual process proposed by the traditional view of evolution. In addition to his scholarly works, Gould has published numerous popular books on paleoanthropology, Darwinian theory, and evolutionary biology.

Grant, Peter, and Rosemary – Biologists whose long-term research focuses on finches in the Galapagos Islands, and the evolutionary impact of climatic and environmental changes on their populations. They live part of the year in the Islands and have received honors for their work since they began in 1973.

Graptolite – A small, colonial, often planktonic marine animal that was very abundant in the oceans 300 to 500 million years ago; now extinct.

Greene, Mott – A historian of science who has written extensively about the development of geological thought during the 19th and early 20th centuries, including the development of the theory of continental drift.

Greenhouse Gases – Gases that absorb and reradiate infrared radiation. When present in the atmosphere, these gases contribute to the greenhouse effect, trapping heat near the surface of the planet. On Earth, the main greenhouse gases are carbon dioxide, water vapor, methane, nitrous oxide, ozone, and some halocarbon compounds.

Group Selection – The selection operating between groups of individuals rather than between individuals. It would produce attributes beneficial to a group in competition with other groups rather than attributes beneficial to individuals.

H

Haeckel, Ernst – A German biologist who lived from 1834–1919, Haeckel was the first to divide animals into protozoan (unicellular) and metazoan (multicellular) forms. His notion of recapitulation is no longer accepted.

Haile Selassie, Yohannes – A paleoanthropologist who, while doing field work in Ethiopia for his doctoral dissertation at the University of California, Berkeley, discovered Ardipithecus ramidus kadabba, a bipedal hominid dated at 5.2 million years old.

Half-Life – The amount of time it takes for one-half of the atoms of a radioactive material to decay to a stable form. For example, the half-life of carbon-14 is 5,568 years.

Hamilton, W. D. – A naturalist, explorer, and zoologist who worked in the world of mathematical models, including "Hamilton's Rule," about the spread through a population of a gene for altruistic self-sacrifice. He was also interested in the evolutionary impact of parasites as the key to many outstanding problems left by Darwin, including the baffling riddle of the evolution of sex. This led him to extensive work in computer simulations.

Haploid – The condition of having only one set of genes or chromosomes. In normally diploid organisms such as humans, only the gametes are haploid.

Haplotype – A set of genes at more than one locus inherited by an individual from one of its parents. It is the multi-locus analog of an allele.

Hardy-Weinberg Principle – In population genetics, the idea that if a population experienced no selection, no mutation, no migration, no genetic drift, and random mating, then the frequency of each allele and the frequencies of genotype in the population would remain the same from one generation to the next.

Hardy-Weinberg Ratio – The ratio of genotype frequencies that evolve when mating is random and neither selection nor drift are operating. For two alleles (A and a) with frequencies p and q, there are three genotypes: AA, Aa, and aa. The Hardy-Weinberg ratio for the three is: p2AA: 2pqAa: q2aa. It is the starting point for much of the theory of population genetics.

Harvey, Ralph – A geologist whose work includes the study of geological processes at a range of scales, from the smallest nanometer to broader-scale interpretations of the history experienced by geological materials.

Heavy Metals – Metals with a high relative atomic mass, such as lead, copper, zinc, and mercury. Many of them are toxic.

Hemoglobin – A protein that carries oxygen from the lungs throughout the body.

Heredity – The process by which characteristics are passed from one generation to the next.

Heritability – Broadly, the proportion of variation (more strictly, variance) in a phenotypic character in a population that is due to individual differences in genotypes. Narrowly, it is defined as the proportion of variation (more strictly, variance) in a phenotypic character in a population that is due to individual genetic differences that will be inherited in the offspring.

Heritable – Partly or wholly determined by genes; capable of being passed from an individual to its offspring.

Herrnstein, Richard J. – A professor of psychology and an author of notable books on intelligence and crime. He has primarily done research on human and animal motivational and learning processes. His books include

Psychology and I. Q. in the Meritocracy, and he coauthored (with Charles Murray) The Bell Curve: Intelligence and Class Structure in American Life (1994).

Heterogametic – The sex with two different sex chromosomes (males in mammals, because they are XY). Compare with homogametic.

Heterozygosity – (for most purposes) The proportion of individuals in a population that are heterozygotes.

Heterozygote – An individual having two different alleles at a genetic locus. Compare with homozygote.

Heterozygote Advantage – A condition in which the fitness of a heterozygote is higher than the fitness of either homozygote.

Heterozygous – Having two different alleles for a particular trait. See also heterozygote.

Hill, Andrew – A paleontologist and professor at Yale University. His work with Mary Leakey's team at Laeotoli, Tanzania, in the 1970s helped lead to the discovery of the fossilized footprints of early hominids and other mammals. His current research interests include hominid evolution, paleoecology, and taphonomy.

HIV – Human Immunodeficiency Virus. The virus causes AIDS by inactivating the T cells of the immune system.

Ho, David – A physician and world-renowned AIDS researcher. Dr. Ho overturned an earlier conventional assumption that the HIV virus remains dormant for up to 10 years in a person before its outbreak into AIDS. His recognition that the virus is active right from the beginning of infection led him to initiate the deployment of a combination of drugs to overpower the virus.

Homeobox – These are relatively short (approximately 180 base pair) sequences of DNA, characteristic of some homeotic genes (which play a central role in controlling body development). Homeoboxes code for a protein "homeodomain," a protein domain that binds to DNA, and can regulate the expression of other genes. These homeodomain motifs are involved in orchestrating the development of a wide range of organisms.

Homeobox Genes – "Homeobox genes" are genes that contain "homeoboxes," and can regulate the expression of other genes. There are at least 24 homeobox genes, some but not all of which are also homeotic in their effect. In general, "homeotic" genes are genes that control the identity of body parts. They are active in the early stages of embryonic development of organisms. Some, but not all, homeotic genes are homeobox genes. "Hox" genes are a subgroup of homeotic homeobox genes that determine positional cell differentiation and development. They lay out the head-to-tail body pattern in very early embryos. The Hox genes are very ancient and widely shared among bilateral animals. After the head-to-tail pattern is established, homeotic genes direct the developmental fates of particular groups of cells.

Homeostasis (Developmental) – A self-regulating process in development, such that the organism grows up to have much the same form independent of the external influences it experiences while growing up.

Homeotic Mutation – A mutation causing one structure of an organism to grow in the place appropriate to another. For example, in the mutation called "antennapedia" in the fruit fly, a foot grows in the antennal socket.

Hominids – Members of the family Hominidae, which includes only modern humans and their ancestors since the human lineage split from the apes.

Homo Erectus – A species of hominid that lived between 1.8 mya and 3,00,000 years ago; the first Homo species to migrate beyond Africa.

Homo Habilis – A species of hominid that lived between 1.9 and 1.8 mya, the first species in genus Homo, and the first hominid associated with clear evidence of tool manufacture and use.

Homo Neanderthalensis – A species of hominid that lived between 1,50,000 and 30,000 years ago in Europe and Western Asia, originally thought to be a geographic variant of Homo sapiens but now generally accepted to be a distinct species.

Homo sapiens – Modern humans, which evolved to their present form about 1,00,000 years ago.

Homogametic – The sex with two of the same kind of chromosomes (females in mammals, because they are XX). Compare with heterogametic.

Homologous Structures – The structures shared by a set of related species because they have been inherited, with or without modification, from their common ancestor. For example, the bones that support a bat's wing are similar to those of a human arm.

Homology – A character shared by a set of species and present in their common ancestor. Compare with analogy. (Some molecular biologists, when comparing two sequences, call the corresponding sites "homologous" if they have the same nucleotide, regardless of whether the similarity is evolutionarily shared from a common ancestor or convergent. They likewise talk about the percent homology between the two sequences. Homology in this context simply means similarity. This usage is frowned upon by many evolutionary biologists but is established in much of the molecular literature.)

Homozygote – An individual having two copies of the same allele at a genetic locus. Also sometimes applied to larger genetic entities, such as a whole chromosome; a homozygote is then an individual having two copies of the same chromosome.

Homozygous – Having identical alleles for a particular trait. See also homozygote.

Horsetail – A seedless plant related to ferns. Around 25 species of only one genus, Equisetum, remain today, whereas many different species, some the size of modern trees, were abundant in ancient swamps. Along with lycophytes and ferns, horsetails were among the first terrestrial plants to appear.

Hox Genes – A particular subgroup of homeobox genes that function to pattern the axis of an organismÕs body and determine where limbs and other body segments will grow as the embryo develops.

Huxley, Thomas Henry – British intellect, photographer, and contemporary of Darwin. He was the first to apply the theory of natural selection to humanity to explain the course of human evolution.

Hybrid – The offspring of a cross between two species.

Hypothesis – An explanation of one or more phenomena in nature that can be tested by observations, experiments, or both. In order to be considered scientific, a hypothesis must be falsifiable, which means that it can be proven to be incorrect.

I

Idealism – The philosophical theory that there is fundamental non-material "ideas," "plans," or "forms" underlying the phenomena we observe in nature. It has been historically influential in classification.

Immigration – The movement of organisms into an area.

Immutability – The ability to withstand change.

Induction – The process of deriving general principles from particular facts.

Inference – A conclusion drawn from evidence.

Inheritance of Acquired Characters – Historically influential but factually erroneous theory that an individual inherits characters that its parents acquired during their lifetimes.

Insectivorous – Feeding largely or exclusively on insects.

Intelligent Design – The non-scientific argument that complex biological structures have been designed by an unidentified supernatural or extraterrestrial intelligence.

Intron – The nucleotide sequences of some genes consist of parts that code for amino acids, and other parts that do not code for amino acids interspersed among them. The interspersed non-coding parts, which are not translated, are called introns; the coding parts are called exons.

Inversion – An event (or the product of the event) in which a sequence of nucleotides in the DNA is reversed or inverted. Sometimes inversions are visible in the structure of the chromosomes.

IQ – An abbreviation of "intelligence quotient," usually defined as the mental age of an individual (as measured by standardized tests) divided by his or her real age and multiplied by 100. This formulation establishes the average IQ as 100. The usefulness and reliability of IQ as a measure of intelligence has been questioned, in part because of the difficulty of devising standardized tests that are free of cultural biases.

Isolating Mechanism – Any mechanism, such as a difference between species in courtship behavior or breeding season, which results in reproductive isolation between the species.

Isolation – Synonym for reproductive isolation.

Isotope – An atom that shares the same atomic number and position as other atoms in an element but has a different number of neutrons and thus a different atomic mass.

J

Jablonski, David – Paleontologist and professor in the Department of the Geophysical Sciences of the University of Chicago. His research emphasizes combining data from living and fossil organisms to study the origins and fates of lineages and adaptations, to develop an understanding of the underlying dynamics of speciation and extinction that could lead to a general theory of evolutionary novelty. He is interested in the way evolutionary patterns are shaped by the alternation of extinction regimes, with rare but influential mass extinctions driving unexpected evolutionary shifts.

Johanson, Don – A paleontologist and founder of the Institute for Human Origins. Johanson discovered Lucy (at that time the oldest, most complete hominid skeleton known) in 1974, and the following year unearthed the fossilized remains of 13 early hominids in Ethiopia. He is the author of several popular books on human origins.

Johnson, Jerry – Johnson's research interests focus on the interactions and evolutionary relationships of amphibian and reptilian species of tropical American and Mexican desert ecosystems. Johnson specializes in field research in places such as Yucatan, Jalisco, Zacatecas, and Chiapas, Mexico. He has done research on the biochemical analysis of rattlesnake venom using immunological techniques, snake ecology, and lizard ecology.

Johnston, Victor – Professor of biopsychology at New Mexico State University in Las Cruces. His research interests include the evolution of consciousness and perceptions of beauty. He is the author of Why We Feel: The Science of Human Emotions.

K

Kegl, Judy – A linguist who works on theoretical linguistics as it applies to signed and spoken languages. Among her research interests is the study of Nicaraguan Sign Language.

Kimbel, Bill – An anatomist, Kimbel worked with Don Johanson and assembled Lucy's skull fragments. In 1991, Kimbel and Yoel Rak found a 70% complete skeleton of Australopithecus afarensis.

Kingdom – The second highest level of taxonomic classification of organisms (below domains). Classification schemes at the kingdom level have changed over time. Recent molecular data have generally reinforced the evolutionary significance of the kingdoms Animalia, Plantae, and Fungi. The single-celled eukaryotes once lumped into the kingdom Protista are now known to be very diverse, and not closely related to one another. The prokaryotic organisms once lumped into the kingdom Monera are now considered to belong to separate domains: Eubacteria and Archaea. See taxon.

Kirchweger, Gina – An Austrian biologist interested in the biological evolution of skin tone. Her essay, "The Biology of Skin Color," concerns the evolution of race.

Kluger, Matthew – A researcher whose work on lizards demonstrated that fever is beneficial and can improve the immune response to infection. The implication for humans is still being researched, but evidence indicates that mild fevers can have a number of important immunological functions that allow us to better fight bacterial and viral infections.

Knowlton, Nancy – Dr. Knowlton is a professor of marine biology at the Scripps Institution of Oceanography, University of California San Diego, and a staff scientist at the Smithsonian Tropical Research Institute in Panama. Her primary research interests concern various facets of marine biodiversity. These include the nature of species boundaries in corals, elucidating biogeographic patterns in tropical seas, the ecology of coral-algal symbiosis, and threshold effects in coral reef ecosystems.

Kondrashov, Alexey – A population geneticist specializing in mathematical analysis who has studied the evolutionary role of slightly deleterious mutations. He has theorized that a primitive organism's strategy for protecting itself against damaging mutations may have been the first step in the evolution of sexual reproduction.

Kreiswirth, Barry – Director of the Public Health Research Institute TB Center in New York, Dr. Kreiswirth uses DNA fingerprinting to study the evolution of antibiotic resistance in Mycobacterium tuberculosis, the pathogen that causes TB.

L

Lamarck, Jean – An 18th-century naturalist, zoologist, and botanist noted for his study and classification of invertebrates, as well as his evolutionary theories. He traveled extensively throughout Europe and was elected to the Academy of Sciences, where he introduced the principles of heredity and acquired characteristics.

Lamarckian Inheritance – Historically, misleading synonym for inheritance of acquired characteristics.

Land Bridge – A connection between two land masses, especially continents (e.g., the Bering land bridge linking Alaska and Siberia across the Bering Strait) that allows migration of plants and animals from one land mass to the other. Before the widespread acceptance of continental drift, the existence of former land bridges was often invoked to explain faunal and floral similarities between continents now widely separated. On a smaller scale, the term may be applied to land connections that have now been removed by recent tectonics or sea-level changes (e.g., between northern France and southeastern England).

Larva (and Larval Stage) – The prereproductive stage of many animals.

The term is particularly apt when the immature stage has a different form from the adult. For example, a caterpillar is the larval stage of a butterfly or moth.

Law – A description of how a natural phenomenon will occur under certain circumstances.

Leakey, Maeve – A paleoanthropologist at the National Museums of Kenya, Maeve is the discoverer of *Kenyanthropus platyops* and *Australopithecus anamensis*. She is married to Richard Leakey.

Leakey, Mary – A British paleoanthropologist described as "a real fossil hunter" and "the real scientist in the family." Her discoveries, some in collaboration with her husband Louis Leakey, included the 1.75-million-year-old skull which first showed the antiquity of hominids in Africa, jaws and teeth of an early Homo species, and fossilized footprints of bipedal hominids.

Leakey, Richard – The son of renowned anthropologists Louis and Mary Leakey, Richard continued their work on early hominids from 1964 until the 1980s, making a number of significant fossils finds in the Lake Turkana area and serving as director of the National Museum of Kenya. Later he devoted his energies to conservation and politics.

Lee, Melanie – A molecular geneticist and microbial biologist, who in the 1980s collaborated with Paul Nurse on novel research that demonstrated the commonality of the genetic code between yeasts and humans. Dr. Lee later took her molecular skills into the pharmaceutical industry and was a leader in moving pharmacology away from animal models and toward the use of recombinant DNA technology for screening potential new therapies. She now heads the research division of Celltech, an international biopharmaceutical company, where her team works on drug discovery and development of new therapies, mainly for the treatment of inflammatory and immune diseases.

Lek – An area of ground divided into territories that are defended by males for the purpose of displaying to potential mates during the breeding season. This form of mating behavior is known as lekking and occurs in various bird species (e.g., the peacock) and also in some mammals. The dominant males occupy the territories at the center of the lek, where they are most likely to attract and mate with visiting females. The outer territories are occupied by subordinate males, who have less mating success. Over successive breeding seasons, younger subordinate males tend to gradually displace older individuals from the most desirable territories and become dominant themselves. The lek territories do not contain resources of value to the female, such as food or nesting materials, although males of some species may build structures such as bowers that form part of their display.

Lemur – A small, tree-dwelling primate that belongs to the group called prosimians.

Lethal Recessive – The case in which inheriting two recessive alleles of a gene causes the death of the organism.

Levine, Michael – Professor of genetics and development in the Molecular and Cell Biology Department at the University of California, Berkeley. Levine was the discoverer (with Bill McGinnis) of homeobox sequences in the homeotic genes Antennapedia and Ultrabithorax while a postdoctoral researcher with Walter Gehring at the University of Basel, Switzerland. His current research involves analysis of gene regulation and patterning in the early Drosophila embryo; studies of embryonic development in the tunicate, *Ciona intestinalis*, focused on the specification of the notochord and tail muscles; and a critical test of classical models for the evolutionary origins of the chordate body plan.

Lineage – An ancestor-descendant sequence of (1) populations, (2) cells, or (3) genes.

Linkage Disequilibrium – A condition in which the haplotype frequencies in a population deviate from the values they would have if the genes at each locus were combined at random. (When no deviation exists, the population is said to be in linkage equilibrium.)

Linked – Of genes, present on the same chromosome.

Linnaean Classification – A hierarchical method of naming classificatory groups, invented by the 18th-century Swedish naturalist Carl von Linné, or Linnaeus. Each individual is assigned to a species, genus, family, order, class, phylum, and kingdom, and some intermediate classificatory levels. Species are referred to by a Linnaean binomial of its genus and species, such as Magnolia grandiflora.

Lively, Curtis – A professor of biology who studies population biology and the ecology and evolution of host-parasite interactions. His laboratory is involved in detailed studies of the interaction between a parasitic trematode and a freshwater New Zealand snail in which both sexual and asexual females coexist.

Locus – The location in the DNA occupied by a particular gene.

Lovejoy, Owen – A paleoanthropologist and consulting forensic anatomist, Lovejoy is known for his analysis of early hominid fossils. His research includes work on Lucy (Australopithecus afarensis).

Lycophyte – Commonly known as club mosses, lycophytes were among the first seedless plants to appear on Earth. Along with horsetails and ferns, these made the planet more hospitable for the first animals.

Lyell, Charles – A 19th-century scientist considered a father of modern geology. Lyell proposed that the geology of Earth is shaped by gradual processes, such as erosion and sedimentation. Lyell's ideas, expressed in his landmark work, Principles of Geology, greatly influenced the young Charles Darwin. Darwin and Lyell later became close friends. While Lyell initially opposed the idea of evolution, he came to accept it after Darwin published On the Origin of Species.

Lyell's Notion of Gradual Change – Also called uniformitarianism, Lyell's notion was that Earth has been shaped by the same forces and processes that operate today, acting continuously over very long periods of time. For example, the ongoing erosion caused by flowing water in a river could, given enough time, carve out the Grand Canyon.

M

Macroevolution – A vague term generally used to refer to evolution on a grand scale, or over long periods of time. There is no precise scientific definition for this term, but it is often used to refer to the emergence or modification of taxa at or above the genus level. The origin or adaptive radiation of a higher taxon, such as vertebrates, could be called a macroevolutionary event.

Macromutation – Mutation of a large phenotypic effect, one that produces a phenotype well outside the range of variation previously existing in the population.

Malaria – A sometimes-fatal disease transferred to humans by mosquitoes, infecting the bloodstream.

Malthus, Thomas – A British economist and demographer best known for his treatise on population growth, which states that people will always threaten to outrun the food supply unless reproduction is closely monitored. His theory was in opposition to the utopians of the 18th century.

Mammals – The group (specifically, a class) of animals, descended from a common ancestor, that share the derived characters of hair or fur, mammary glands, and several distinctive features of skeletal anatomy, including a particular type of middle ear. Humans, cows, and dolphins are all mammals.

Mammary Glands – Only found in mammals, these are specialized glands that can produce milk for feeding young.

Mandible – A part of the bony structure of a jaw. In vertebrates, it is the lower jaw; in birds, the lower bill; in arthropods, one of the paired appendages closest to the mouth.

Margulis, Lynn – A biologist who developed the serial endosymbiosis theory of origin of the eukaryotic cell. Although now accepted as a plausible theory, both she and her theory were ridiculed by mainstream biologists for a number of years.

Marsupial Mammals – A group (specifically, an order) of mammals whose females give birth to young at a very early stage of development. These newborns complete their development while sucking in a pouch, which is a permanent feature in the female. Examples include kangaroos and opossums.

Mastodon – An extinct elephant-like mammal.

Mayr, Ernst – Mayr's work has contributed to the synthesis of Mendelian genetics and Darwinian evolution, and to the development of the biological species concept. Mayr has been universally recognized and acknowledged as one of the leading evolutionary biologists of the 20th century.

McGinnis, William – A professor of biology at the University of California, San Diego. Discover (with Mike Levine) of homeoboxes, the sequences of DNA that are characteristic of homeotic genes, which play a central role in specifying body development. His current research uses both genetics and biochemistry to examine such questions as how molecular variations in the Hox genes that specify the head-tail pattern of an organism can generate variety in animal shapes during evolution, and what the molecular changes were that allowed single-celled animals to become multicellular.

Meiosis – A special kind of cell division that occurs during the reproduction of diploid organisms to produce the gametes. The double set of genes and chromosomes of the normal diploid cells is reduced during meiosis to a single haploid set in the gametes. Crossing-over and therefore recombination occurs during a phase of meiosis.

Meme – The word coined by Richard Dawkins for a unit of culture, such as an idea, skill, story, or custom, passed from one person to another by imitation or teaching. Some theorists argue that memes are the cultural equivalent of genes, and reproduce, mutate, are selected, and evolve in a similar way.

Mendel, Gregor – An Austrian monk whose plant breeding experiments, begun in 1856, led to insights into the mechanisms of heredity that are the foundation of genetics today. His work was ignored in his lifetime and only rediscovered in 1900. See Mendelian inheritance.

Mendelian Inheritance – The mode of inheritance of all diploid species, and therefore of nearly all multicellular organisms. Inheritance is controlled by genes, which are passed on to the offspring in the same form as they were inherited from the previous generation. At each locus, an individual has two genes – one inherited from its father and the other from its mother. The two genes are represented in equal proportions in its gametes.

Messenger RNA (mRNA) – A kind of RNA produced by transcription from the DNA, and which acts as the message that is decoded to form proteins.

Metabolism – The chemical processes that occur in a living organism in order to maintain life. There are two kinds of metabolism: constructive metabolism, or anabolism, the synthesis of the proteins, carbohydrates, and fats which form tissue and store energy; and destructive metabolism, or catabolism, the breakdown of complex substances, producing energy and waste matter.

Metamorphosis – One or more changes in form during the life cycle of an organism, such as an amphibian or insect, in which the juvenile stages differ from the adult. An example is the transition from a tadpole to an adult frog. The term "complete metamorphosis" is applied to insects such as butterflies in which the caterpillar stage is distinct from the adult. "Incomplete metamorphosis" describes the life histories of insects such as locusts in which the young go through a series of larval stages, each of which bears similarities to the adult. Metamorphosis in both insects and amphibians is controlled by hormones, and often involves considerable destruction of larval tissues by enzymes.

Metazoans – All animals that are multicellular and whose cells are organized into tissues and organs. In the simplest metazoans, only an inner and outer layer can be distinguished.

Microbe – A nonscientific and very general term, with no taxonomic significance, sometimes used to refer to microscopic (not visible to the unaided eye) organisms. The term often refers to bacteria or viruses that cause disease or infection.

Microevolution – Evolutionary changes on a small scale, such as changes in gene frequencies within a population.

Miller, Geoffrey – Author of The Mating Mind, Miller is known for his research on evolutionary psychology and sexual selection. He believes that our minds evolved not only as survival machines, but also as courtship machines – at least in part, to help us attract a mate and pass on genes.

Miller, Ken – A cell biologist and professor of biology at Brown University. Miller's academic research focuses on the structure and function of biological membranes. He is the coauthor of widely used high school and college biology textbooks, and he has also written Finding Darwin's God: A Scientist's Search for Common Ground Between God and Evolution.

Miller, Veronica – A German virologist whose research has focused on HIV-AIDS. Miller was the first researcher to announce that an interruption in drug treatment among AIDS patients may result in reversion of drug-resistant virus to its wild type. This led other researchers and clinicians to explore "structured treatment interruptions" among some patients as an experimental treatment option.

Mimicry – A case in which one species looks more or less similar to another species. See Batesian mimicry and Müllerian mimicry.

Mitochondrial DNA – DNA found in the mitochondrion; a small round body found in most cells. Because mitochondria are generally carried in egg cells but not in sperm, mitochondrial DNA is passed to offspring from mothers, but not fathers.

Mitochondrion – A kind of organelle in eukaryotic cells. Mitochondria produce enzymes to convert food to energy. They contain DNA coding for some mitochondrial proteins.

Mitosis – Cell division. All cell division in multicellular organisms occurs by mitosis except for the special division called meiosis that generates the gametes.

Modern Synthesis – The synthesis of natural selection and Mendelian inheritance. Also called Neo-Darwinism.

Molecular Clock – The theory that molecules evolve at an approximately constant rate. The difference between the form of a molecule in two species is then proportional to the time since the species diverged from a common ancestor, and molecules become of great value in the inference of phylogeny.

Molecular Geneticists – Scientists who study genes and characters through work with the molecules that make up and interact with DNA.

Mollusk – An invertebrate that has a fleshy, muscular body. The phylum Mollusca includes snails, bivalves, squids, and octopuses.

Monkey Trial – In 1925, John Scopes was convicted and fined $100 for teaching evolution in his Dayton, Tenn., classroom in the first highly publicized trial concerning the teaching of evolution. The press reported that although they lost the case, Scopes's team had won the argument. The verdict had a chilling effect on teaching evolution in the classroom, however, and not until the 1960s did it reappear in schoolbooks.

Monogamy – A reproductive strategy in which one male and one female mate and reproduce exclusively with each other. Contrast with polygyny and polyandry.

Monophyletic Group – A set of species containing a common ancestor and all of its descendants, and not containing any organisms that are not the descendants of that common ancestor.

Monotremes – A group (specifically, an order) of mammals whose females lay eggs. The young hatch and continue to develop in the mother's pouch, which is present only when needed. Two species of spiny anteater and the duck-billed platypus are the only living monotremes.

Moore, James – The author, with Adrian Desmond, of an authoritative biography of Charles Darwin, Moore has made a 20-year study of Darwin's life. With degrees in science, divinity, and history, he has taught the history of science at Harvard University and at the Open University in the U.K.

Morphology – The study of the form, shape, and structure of organisms.

Mueller, Ulrich G. – A zoologist and professor whose research aims at understanding microevolutionary forces and macroevolutionary patterns that govern the evolution of organismal interactions, particularly the evolution of mutualisms and the evolution of social conflict and cooperation. Mueller's current research focuses on the coevolution between fungus-growing ants and their fungi and the evolutionary ecology of halictine bees.

Müllerian Mimicry – A kind of mimicry in which two poisonous species evolve to look like one another.

Murray, Charles – An author and policy analyst who has written many controversial and influential books on social policy. He is coauthor with Richard J. Herrnstein of The Bell Curve: Intelligence and Class Structure in American Life (1994). He has also written Losing Ground: American Social Policy 1950–1980 (1984), which argues for the abolishment of the welfare system, The Underclass Revisited (1999), and Income, Inequality, and IQ (1998).

Mutation – A change in genetic material that results from an error in the replication of DNA. Mutations can be beneficial, harmful, or neutral.

N

Nagel, Ronald – A hematologist and professor at Albert Einstein College of Medicine. His research includes molecular, biochemical, and physiological studies of genetic red blood cell defects, including sickle cell.

Natural Selection – The differential survival and reproduction of classes of organisms that differ from one another in one or more usually heritable characteristics. Through this process, the forms of organisms in a population that are best adapted to their local environment increase in frequency relative to less well-adapted forms over a number of generations. This difference in survival and reproduction is not due to chance.

Neanderthal – A hominid, similar to but distinct from modern humans, that lived in Europe and Western Asia about 1,50,000 to 30,000 years ago.

Nelson, Craig – A professor of biology and environmental affairs at Indiana University in Bloomington. His research focuses on evolutionary ecology.

Neo-Darwinism – (1) Darwin's theory of natural selection plus Mendelian inheritance. (2) The larger body of evolutionary thought that was inspired by the unification of natural selection and Mendelism. A synonym of the modern synthesis.

Nervous System – An organ system, composed of a network of cells called neurons, that allows an animal to monitor its internal and external environment, and to move voluntarily or in response to stimulation.

Neural – Related to nerves and neurons.

Neutral Drift – Synonym of genetic drift.

Neutral Mutation – A mutation with the same fitness as the other allele or alleles at its locus.

Neutral Theory (and Neutralism) – The theory that much evolution at the molecular level occurs by genetic drift.

Newton, Isaac – An English physician and mathematician, considered the culminating figure of the scientific revolution of the 17th century. He is best known for his explanation of gravity and for laying the foundation for modern physical optics.

Niche – The ecological role of a species; the set of resources it consumes and the habitats it occupies.

Nilsson, Dan-Erik – Professor Nilsson heads the Functional Morphology Division of the Department of Zoology at Lund University in Sweden. His main research interest is the optics and evolution of invertebrate eyes.

Nilsson, Lennart – A Swedish photographer who began as a photojournalist, Nilsson soon began exploring new techniques such as the use of endoscopes and electron microscopes to photograph the inner mysteries of the human body. He published a book entitled A Child is Born of his photographs of the beginning of life, and made a number of films, including the mini-series Odyssey of Life, a coproduction between WGBH/NOVA and SVT Swedish Television.

Nitrogen Fixation – A chemical process by which nitrogen in the atmosphere is assimilated into organic compounds. Only certain bacteria are able to fix atmospheric nitrogen, which then becomes available to other organisms through the food chains.

Nomadic – Having to do with nomads, people who live in no fixed place but move in search of food or grazing land for their animals, of a wandering lifestyle.

Notochord – A flexible skeletal rod running the length of the body in the embryos of the chordates (including the vertebrates). In some simpler types, such as sea-squirts, only the free-swimming larva has a notochord; in others, such as the lancelets and lampreys, the notochord remains the main axial support, and in vertebrates, it is incorporated into the backbone as the embryo develops.

Novacek, Michael J. – Paleontologist with the American Museum of Natural History. Dr. Novacek's research interests include the evolution of and relationships among organisms, particularly mammals. He is the author of Dinosaurs of the Flaming Cliffs, an account of AMNH's Gobi Desert expeditions.

Nucleotide – A unit building block of DNA and RNA. A nucleotide consists of a sugar and phosphate backbone with a base attached.

Nucleus – A region of eukaryotic cells, enclosed within a membrane, containing the DNA.

Numerical Taxonomy – In general, any method of taxonomy using numerical measurements. In particular, it often refers to phenetic classification using large numbers of quantitatively measured characters.

Nurse, Paul – A pioneer in genetic and molecular studies who revealed the universal machinery for regulating cell division in all eukaryotic organisms, from yeasts to frogs to human beings.

O

O'Brien, Stephen J. – A geneticist at the National Cancer Institute whose research interests include the evolutionary history of the immunological response in mammals to retroviruses like HIV. With his colleagues, he discovered a mutation that can protect individuals from infection by HIV, the virus that causes AIDS.

Order – The taxonomic classification level between class and family. For example, within the class Mammalia, there are several orders, including the meat-eaters, who make up the order Carnivora; and the insect-eaters, grouped together in the order Insectivora. The orders in turn are divided into families; the order Carnivora includes the families Felidae (the cats), Canidae (the dogs), and Ursidae (the bears), among others. See also taxon.

Organelle – Any of a number of distinct small structures found in the cytoplasm (and therefore outside the nucleus) of eukaryotic cells (e.g., mitochondrion and chloroplast).

Organisms – Living things.

Orthogenesis – The erroneous idea that species tend to evolve in a fixed direction because of some inherent force driving them to do so.

Owen, Richard – A 19th-century British comparative anatomist, who coined the word "dinosaur" to describe a breed of large, extinct reptiles. He was the first to propose that dinosaurs were a separate taxonomic group. Owen opposed Darwin's theory of evolution, but ultimately his work helped support evolutionary arguments.

Ozone Layer – The region of the atmosphere, generally 11–26 km (7–16 miles) above Earth, where ozone forms in high concentrations. The ozone layer absorbs ultraviolet radiation, shielding Earth from its damaging effects.

P

Paleoanthropologist – A scientist who uses fossil evidence to study early human ancestors.

Paleobiology – The biological study of fossils.

Paleontologist – A scientist who studies fossils to better understand life in prehistoric times.

Paleontology – The scientific study of fossils.

Pangaea – A supercontinent that began to break apart into the modern continents about 260 million years ago, causing the isolation (and separate evolution) of various groups of organisms from each other.

Panmixis – Random mating throughout a population.

Paradox – A seemingly absurd or contradictory, though often true, statement.

Parapatric Speciation – Speciation in which the new species forms from a population contiguous with the ancestral species' geographic range.

Paraphyletic Group – A set of species containing an ancestral species together with some, but not all, of its descendants. The species included in the group are those that have continued to resemble the ancestor; the excluded species have evolved rapidly and no longer resemble their ancestor.

Parasite – An organism that lives on or in a plant or animal of a different species, taking nutrients without providing any benefit to the host.

Parish, Amy – A biological anthropologist and primatologist whose research focuses on the social behavior of bonobos ("pygmy chimpanzees," or *Pan paniscus*). In addition to comparative work with chimpanzees (*Pan troglodytes*) and endocrinological investigations, Dr. Parish studies reciprocity in chimpanzees, bonobos, and hunter-gatherers.

Parsimony – The principle of phylogenetic reconstruction in which the phylogeny of a group of species is inferred to be the branching pattern requiring the smallest number of evolutionary changes.

Parthenogenesis – Development from an egg cell that has not been fertilized. The term for a certain form of asexual reproduction that is found in some lizards, insects (notably among aphids), and certain other organisms.

Particulate – (as property of theory of inheritance) A synonym of atomistic.

Paternity – The identity of the father of an offspring.

Pathogen – A microorganism that causes disease.

Pathological – Related to or caused by disease.

Penicillin – The first antibiotic discovered; penicillin is derived from the mold *Penicillium notatum*. It is active against a wide variety of bacteria, acting by disrupting the synthesis of the bacterial cell wall.

Peripatric Speciation – A synonym of peripheral isolate speciation.

Peripheral Isolate Speciation – A form of allopatric speciation in which the new species is formed from a small population isolated at the edge of the ancestral population's geographic range. Also called peripatric speciation.

Pesticide-Resistant Insects – Insects with the ability to survive and reproduce in the presence of pesticides. These resistant variants increase in frequency over time if pesticides remain present in their environment.

Petrie, Marion – A behavioral ecologist at the University of Newcastle in the U.K., Dr. Petrie's research interests include the links between sexual selection and speciation, and how males and females assess genetic quality in a mate.

Phenetic Classification – A method of classification in which species are grouped together with other species that they most closely resemble phenotypically.

Phenetic Species Concept – A concept of species according to which a species is a set of organisms that are phenotypically similar to one another. Compare with biological species concept, cladistic species concept, ecological species concept, and recognition species concept.

Phenotype – The physical or functional characteristics of an organism, produced by the interaction of genotype and environment during growth and development.

Phenotypic Characters – Individual traits that can be observed in an organism (including appearance and behavior) and that result from the interaction between the organism's genetic makeup and its environment.

Pheromone – A chemical substance produced by some organisms and emitted into the environment to communicate with others of the same species. Pheromones play an important role in the social behavior of certain animals, especially insects and some mammals. They are used to mark out territories, to attract mates, to lay trails, and to promote social cohesion and coordination in colonies. Examples are the sex attractants secreted by moths to attract mates and the queen substance produced by queen honeybees, which controls the development and behavior of worker bees. Pheromones are usually volatile organic molecules that are effective at very low concentrations, as little as one part per million.

Photoreceptor Cell – A cell, functionally part of the nervous system, that reacts to the presence of light. It usually contains a pigment that undergoes a chemical change when light is absorbed. This chemical change stimulates electrical changes in the photoreceptor that, when passed along and processed by other neurons, form the basis of vision.

Photosynthesis – The fundamental biological process by which green plants make organic compounds such as carbohydrates from atmospheric carbon dioxide and water using light energy from the Sun. The process has two main phases: the light-dependent light reaction responsible for the initial capture of energy, and the light-independent dark reaction in which this energy is stored in the chemical bonds of organic molecules. Because virtually all other forms of life are directly or indirectly dependent on green plants for food, photosynthesis is the basis for almost all life on Earth.

Phylogeny – The study of ancestral relations among species, often illustrated with a "tree of life" branching diagram, which is also known as a phylogenetic tree.

Phylum (Plural Phyla) – One of the highest levels of taxonomic classification. See taxon.

Phytoplankton – Microscopic aquatic organisms that, like plants, use photosynthesis to capture and harness solar energy.

Pickford, Martin – A paleontologist at the College de France in Paris. In 2000, Pickford and Brigitte Senut discovered Orrorintugensis, a proto-hominid dated at 6 million years old.

Pinker, Steven – A psychologist and professor with a special interest in language, linguistic behavior, and cognitive science. Pinker's publications include the popular science books The Language Instinct and How the Mind Works.

Placental Mammals – A group (specifically, an order) of mammals in which the young develop inside the mother, attached to her, and nourished by a specialized structure called the placenta. In placental mammals, the young are born in an advanced stage of development. Compare with marsupial and monotreme.

Placoderm – An extinct bottom-dwelling fish that was among the first to develop jaws and paired fins.

Plan of Nature – The philosophical theory that nature is organized according to a plan. It has been influential in classification and is a kind of idealism.

Plankton – Minute or microscopic animals (zooplankton) and plants (phytoplankton) that float and drift in water, usually near the surface. In the top meter or two of water, both in the sea and in freshwater, small plants can photosynthesize, and abundant microscopic life can be observed. Many organisms that are sessile (attached to a surface) as adults disperse by means of a planktonic larval stage.

Plasmid – A genetic element that exists (or can exist) independently of the main DNA in the cell. In bacteria, plasmids can exist as small loops of DNA and be passed between cells independently.

Plate Tectonics – The theory that the surface of Earth is made of a number of plates, which have moved throughout geological time to create the present-day positions of the continents. Plate tectonics explains the location of mountain building, as well as earthquakes and volcanoes. The rigid plates consist of continental and oceanic crust together with the upper mantle, which "float" on the semi-molten layer of the mantle beneath them and move relative to each other across the planet. Six major plates (Eurasian, American, African, Pacific, Indian, and Antarctic) are recognized, together with a number of smaller ones. The plate margins coincide with zones of seismic and volcanic activity.

Poisson Distribution – A frequency distribution for the number of events per unit time, when the number of events is determined randomly, and the probability of each event is low.

Polyandry – A reproductive system in which one female mates with many males. Seahorses and jacanas are examples of polyandrous species, which are less common than monogamous or polygynous species.

Polygyny – Reproductive strategy in which one male mates with several females. Lions, peacocks, and gorillas all have polygynous mating systems. Compare with polyandry and monogamy.

Polymorphism – A condition in which a population possesses more than one allele at a locus. Sometimes it is defined as the condition of having more than one allele with a frequency of more than 5% in the population.

Polyphyletic Group – A set of species descended from more than one common ancestor. The ultimate common ancestor of all species in the group is not a member of the polyphyletic group.

Polyploid – An individual containing more than two sets of genes and chromosomes.

Population – A group of organisms, usually a group of sexual organisms that interbreed and share a gene pool.

Population Genetics – The study of processes influencing gene frequencies.

Postulate – A basic principle.

Postzygotic Isolation – A form of reproductive isolation in which a zygote is successfully formed but then either fails to develop or develops into a sterile adult. Donkeys and horses are postzygotically isolated from one another; a male donkey and a female horse can mate to produce a mule, but the mule is sterile.

Prezygotic Isolation – A form of reproductive isolation in which the two species never reach the stage of successful mating, and thus no zygote is formed. Examples would be species that have different breeding seasons or courtship displays, and which therefore never recognize one another as potential mates.

Primate – A mammal belonging to the order Primates (about 195 species), which includes prosimians, monkeys, apes, and humans. Primates probably evolved from insectivorous climbing creatures like tree shrews and have many adaptations for climbing, including five fingers and five toes with opposable first digits (except in the hind feet of humans). They have well-developed hearing and sight, with forward-facing eyes allowing binocular vision, and large brains. The young are usually produced singly and undergo a long period of growth and development to the adult form. Most primates are arboreal, but the great apes and humans are largely terrestrial.

Prokaryote – A cell without a distinct nucleus. Bacteria and some other simple organisms are prokaryotic. Compare with eukaryote. In classificatory terms, the group of all prokaryotes is paraphyletic.

Proofreading Enzymes – Mistakes during DNA replication can be recognized and repaired by proofreading enzymes. A mismatched nucleotide may occur at the rate of one per 1,00,000 base pairs, causing a pause in replication. DNA repair enzymes perform a proofreading function and reduce the error rate to one per billion. Many complex mechanisms have evolved to repair damage and alterations in DNA, which can occur as a result of damage from ultraviolet radiation, X-rays, etc., as well as from errors during replication. (Over 50 genes have been found in yeast to be involved in repair mechanisms).

Prosimian – One of the groups of primates that includes lemurs and lorises; the other two primate groups are tarsoids and anthropoids.

Protein – A molecule made up of a sequence of amino acids. Many of the important molecules in a living thing – for example, all enzymes – are proteins.

Protozoa – A group of unicellular, usually microscopic, organisms now classified in various phyla of the kingdom Protoctista. They were formerly regarded either as a phylum of simple animals or as members of the kingdom Protista. Most feed on decomposing dead organic matter, but some are parasites, including the agents causing malaria (Plasmodium) and sleeping sickness (Trypanosoma), and a few contain chlorophyll and carry out photosynthesis, like plants.

Pseudogene – A sequence of nucleotides in the DNA that resembles a gene but is nonfunctional for some reason.

Pupa (Plural Pupae) – The third stage of development in the life cycle of some insects, including flies, butterflies (in which it is the chrysalis), ants, bees, and beetles. During the pupal stage locomotion and feeding cease and metamorphosis from the larva to the adult form takes place. The adult emerges by cutting or digesting the pupal case after a few days or several months.

Purine – A kind of base in the DNA; adenine (A) and guanine (G) are purines.

Pyrimidine – A kind of base. In DNA, cytosine (C) and thymine (T) are pyrimidines. In RNA, cytosine (C) and uracil (U) are pyrimidines.

Q

Quantitative Character – A character showing continuous variation in a population.

R

Radioactivity – The emission of energy due to changes in the nucleus of an atom. Such spontaneously released radiation is a characteristic of certain elements and at some levels can be harmful.

Radiometric Dating – A dating technique that uses the decay rate of radioactive isotopes to estimate the age of an object.

Rak, Yoel – An Israeli paleoanthropologist and anatomist whose research interests include facial morphology of fossil hominids. Rak was part of the team that found a 2.3-million-year-old skull fragment from the genus Homo at Hadar, Ethiopia.

Random Drift – Synonym of genetic drift.

Random Mating – A mating pattern in which the probability of mating with another individual of a particular genotype (or phenotype) equals the frequency of that genotype (or phenotype) in the population.

Recanted – Withdrew a statement or opinion; disavowed a former assertion.

Recapitulation – A partly or wholly erroneous hypothesis stating that an individual, during its development, passes through a series of stages corresponding to its successive evolutionary ancestors. According to the recapitulation hypothesis, an individual thus develops by "climbing up its family tree."

Receptors – Proteins that can bind to other specific molecules. Usually on the surface of a cell, receptors often bind to antibodies or hormones.

Recessive – An allele (A) is recessive if the phenotype of the heterozygote (Aa) is the same as the homozygote (aa) for the alternative allele (a) and different from the homozygote for the recessive (AA). The allele (a) controls the heterozygote's phenotype and is called dominant. An allele may be partly, rather than fully, recessive; in that case, the heterozygous phenotype is nearer to, rather than identical with, the homozygote for the dominant allele.

Recognition Species Concept – A concept of species according to which a species is a set of organisms that recognize one another as potential mates; they have a shared mate recognition system. Compare with biological species concept, cladistic species concept, ecological species concept, and phenetic species concept.

Recombination – An event, occurring by the crossing-over of chromosomes during meiosis, in which DNA is exchanged between a pair of chromosomes of a pair. Thus, two genes that were previously unlinked, being on different chromosomes, can become linked because of recombination, and linked genes may become unlinked.

Reinforcement – An increase in reproductive isolation between incipient species by natural selection. Natural selection can directly favor only an increase in prezygotic isolation; reinforcement therefore amounts to selection for assortative mating between the incipiently speciating forms.

Relative Dating – The process of ordering fossils, rocks, and geologic events from oldest to youngest. Because of the way sedimentary rocks form, lower layers in most series are older than higher layers, making it possible to determine which fossils found in those layers are oldest and which are youngest. By itself, relative dating cannot assign any absolute age to rocks or fossils.

Reproductive Character Displacement – The increased reproductive isolation between two closely related species when they live in the same geographic region (sympatry) as compared with when they live in separate geographic regions. A kind of character displacement in which the character concerned influences reproductive isolation, not ecological competition.

Reproductive Isolation – Two populations or individuals of opposite sex are considered reproductively isolated from one another if they cannot together produce fertile offspring. See prezygotic isolation and postzygotic isolation.

Retina – The back wall of the eye onto which images are projected. From the retina, the information is sent to the brain via the optic nerve.

Ribosomal RNA (rRNA) – The kind of RNA that constitutes the ribosomes and provides the site for translation.

Ribosome – The site of protein synthesis (or translation) in the cell, mainly consisting of ribosomal RNA.

Ring Species – A situation in which two reproductively isolated populations (see reproductive isolation) living in the same region are connected by a geographic ring of populations that can interbreed.

RNA (Ribonucleic Acid) – Messenger RNA, ribosomal RNA, and transfer RNA are its three main forms. These act as the intermediaries by which the hereditary code of DNA is converted into proteins. In some viruses, RNA is itself the hereditary molecule.

S

Saag, Michael – Dr. Saag is director of the AIDS Outpatient Clinic and associate professor of medicine at the University of Alabama, Birmingham. He is also the associate director for clinical care and therapeutics at the UAB AIDS Center. Dr. Saag's research activities focus on both clinical and basic aspects of the human immunodeficiency virus. He serves on several state and national advisory panels, including the NIH/NIAID AIDS Clinical Trials Group Executive Committee.

Sagittal Crest – A ridge of bone projecting up from the top midline of the skull, running from front to back. It serves as a muscle attachment area for the muscles that extend up the side of the head from the jaw. The presence of a sagittal crest indicates extremely strong jaw muscles.

Schneider, Chris – A biologist and professor at Boston University whose research focuses on the evolution of vertebrate diversity in tropical systems and the scientific basis for conservation of tropical diversity. He uses a variety of molecular genetic methods, such as DNA sequencing, to study speciation, systematics, and biogeography of terrestrial vertebrates, with an emphasis on reptiles and amphibians.

Schultz, Ted R. – An ant systematist at the Smithsonian Institution, Dr. Schultz studies the evolution of the symbiosis between fungus-growing ants and the fungi they cultivate.

Science – A way of knowing about the natural world based on observations and experiments that can be confirmed or disproved by other scientists using accepted scientific techniques.

Scopes, John – The 24-year-old teacher in the public high school in Dayton, Tenn., who was the defendant in the "monkey trial" of 1925. He agreed to be the focus of a test case attacking a newly passed Tennessee state law against teaching evolution or any other theory denying the biblical account of the creation of man, and was arrested and tried, with the American Civil Liberties Union backing his defense.

Scott, Eugenie C. – A human biologist specializing in medical anthropology and skeletal biology. As executive director of the National Center for Science Education, Scott is an advocate of church/state separation in schools, and speaks widely about science, evolution, and natural selection.

Scott, Matthew P. – A professor and researcher whose work in developmental biology explores how homeotic genes orchestrate differentiation and multicellular organization.

Sedimentary Rocks – Rocks composed of sediments, usually with a layered appearance. The sediments are composed of particles that come mostly from the weathering of pre-existing rocks, but often include material of organic origin; they are then transported and deposited by wind, water, or glacial ice. Sedimentary rocks are deposited mainly under water, usually in approximately horizontal layers (beds). Clastic sedimentary rocks are formed from the erosion and deposition of pre-existing rocks and are classified according to the size of the particles. Organically formed sedimentary rocks are derived from the remains of plants and animals (e.g., limestone and coal). Chemically formed sedimentary rocks result from natural chemical processes and include sedimentary iron ores. Many sedimentary rocks contain fossils.

Selection – Synonym of natural selection.

Selectionism – The theory that some classes of evolutionary events, such as molecular or phenotypic changes, have mainly been caused by natural selection.

Selective Pressures – Environmental forces such as scarcity of food or extreme temperatures that result in the survival of only certain organisms with characteristics that provide resistance.

Senut, Brigitte – An anatomist at France's National Museum of Natural History. In 2000, Senut and Martin Pickford discovered Orrorintugensis, a proto-hominid dated at 6 million years old.

Separate Creation – The theory that species have separate origins and never change after their origin. Most versions of the theory of separate creation are religiously inspired and suggest that the origin of species occurs by supernatural action.

Sex Chromosome – The chromosome or chromosomes that influence sex determination. In mammals, including humans, the X and Y chromosomes are the sex chromosomes (females are XX, males XY). Compare with autosome.

Sexual Selection – A selection on mating behavior, either through competition among members of one sex (usually males) for access to members of the other sex or through choice by members of one sex (usually females) of certain members of the other sex. In sexual selection, individuals are favored by their fitness relative to other members of the same sex, whereas natural selection works on the fitness of a genotype relative to the whole population.

Sexually Dimorphic – When males and females of a species have considerably different appearances, which may include size, coloration, or other features, such as special plumage.

Shubin, Neil – A paleontologist who is known for his work on early tetrapods (any creature with four limbs). He presented a hypothesis of general patterns of the development of tetrapod limbs which changed the way scientists think about this field. The study of limbs is crucial to evolutionary science; one example of why this is important is that human development would have been impossible without limbs.

Sickle Cell Anemia – A disease in which poorly formed red blood cells cannot bind correctly to oxygen, resulting in low iron, blood clotting, and joint pain.

Simpson, George Gaylord – One of the most influential paleontologists of the 20th century and a leading developer of the modern synthesis. He wrote hundreds of technical papers in addition to many widely read popular books and textbooks, and was a leading expert on Mesozoic, Paleocene, and South American mammals.

Small, Meredith – A professor of anthropology. Her research interests include primate behavior and ecology; mating strategies; reproduction; and the evolution of human behavior. Small's publications include Female Choices: Sexual Behavior of Female Primates, What's Love Got to Do with It? and The Evolution of Human Mating.

Smith, John Maynard – An eminent evolutionary biologist and author of many books on evolution, both for scientists and the general public. A professor emeritus at the University of Sussex, his research interests include the evolution of human mitochondrial DNA sequences and the investigation of evidence for extensive recombination.

Smith, Tom – An ornithologist and conservation biologist, Smith is executive director of the Center for Tropical Research at San Francisco State University. His work combines basic research in ecology and evolutionary science with applied research in conservation biology. Among other issues, Smith is interested in the role of ecological gradients in speciation and maintaining species diversity.

Social Darwinism – A doctrine that applies the principles of selection to the structure of society, asserting that social structure is determined by how well people are suited to living conditions.

Spacer Region – A sequence of nucleotides in the DNA between coding genes.

Speciation – Changes in related organisms to the point where they are different enough to be considered separate species. This occurs when populations of one species are separated and adapt to their new environment or conditions (physiological, geographic, or behavioral).

Species – An important classificatory category, which can be variously defined by the biological species concept, cladistic species concept, ecological species concept, phenetic species concept, and recognition species concept. The biological species concept, according to which a species is a set of interbreeding organisms, is the most widely used definition, at least by biologists who study vertebrates. A particular species is referred to by a Linnaean binomial, such as Homo sapiens for human beings.

Sponge – A member of the phylum Porifera, marine and freshwater invertebrates that live permanently attached to rocks or other surfaces. The body of a sponge is hollow and consists basically of an aggregation of cells between which there is little nervous coordination, although they do have specialized sets of cells that perform different functions. One set of cells causes water to flow in through openings in the body wall and out through openings at the top; food particles are filtered from the water by these cells. Other cells construct a stiffening skeletal framework of spicules of chalk, silica, or fibrous protein to support the body.

Stabilizing Selection – A form of selection that tends to keep the form of a population constant. Individuals with the mean value for a character have high fitness; those with extreme values have low fitness.

Stepped Cline – A cline with a sudden change in gene or character frequency.

Stromatoporoid – Once thought to be related to the corals, are now recognized as being calcareous sponges. Sponges similar to fossil stromatoporoids are found in the oceans today. Like modern sponges, stromatoporoid created currents to pump water in and out of their body, where they filtered out tiny food particles. Fossil stromatoporoids can be massive, chocolate-drop in shape, tabular, encrusting, cylindrical, or even arm-shaped ("ramose"). There are two main groups of fossil stromatoporoids that lived in different eras, the Paleozoic, and the Mesozoic. After their appearance in the Ordovician, the Paleozoic stromatoporoids were dominant reef builders for over 100 million years. The second group of stromatoporoids, from the Mesozoic, may represent a distinct group with a similar growth form. They were also important contributors to reef formation, especially during the Cretaceous.

Subduction Zone – A zone where rocks of an oceanic plate are forced to plunge below much thicker continental crust, along margins between adjoining plates. As the plate descends it melts and is released into the magma below Earth's crust. Such a zone is marked by volcanoes and earthquakes. See plate tectonics.

Substitution – The evolutionary replacement of one allele by another in a population.

Supernatural – Relating to phenomena that cannot be described by natural laws, cannot be tested by scientific methodology, and are therefore outside the realm of science.

Symbiosis – A relationship of mutual benefit between two organisms that live together.

Sympatric Speciation – Speciation via populations with overlapping geographic ranges.

Sympatry – Living in the same geographic region. Compare with allopatry.

Syntax – The rules by which words are combined to form grammatical sentences.

Systematics – A near synonym of taxonomy.

T

Tarsier – One of three species of small nocturnal primates belonging to the genus Tarsius, found in Sumatra, Borneo, Celebes, and the Philippines. They have a naked tail measuring 130–270 mm (about 5–11 inches) long, which makes up about half the total length of their bodies (220–460 mm, or 8–19 inches). Tarsiers have enormous eyes, large hairless ears, and gripping pads at the end of their digits. They are mainly arboreal, using both hands to seize insects and small vertebrates such as lizards.

Taxon (Plural Taxa) – Any named taxonomic group, such as the family Felidae, or the genus Homo, or the species Homo sapiens. Also, a formally recognized group, as distinct from any other group (such as the group of herbivores, or the group of tree-climbers).

Taxonomy – The theory and practice of biological classification.

Terrestrial – Living on land.

Tetrapod – A member of the group made up of amphibians, reptiles, birds, and mammals.

Thecodont – The thecodonts were a diverse group of Triassic reptiles that included large four-legged carnivores, armored herbivores, small, agile two- and four-legged forms, and crocodile-like aquatic reptiles. They gave rise to crocodiles, dinosaurs, and pterosaurs. The term Thecodontia is no longer used, as they are a paraphyletic group. The thecodonts are therefore an evolutionary grade of animals, rather than a clade. Most paleontologists now use the term "basal archosaur" to refer to thecodonts. As a group, they are defined by certain shared ancestral features, such as teeth in sockets, an archosaurian characteristic that was inherited by the dinosaurs. The name thecodont is actually Latin for "socket-tooth." Members of the group show a general trend toward a more upright, less sprawling stance, with the hindlimbs especially being progressively positioned more directly beneath the body, until some could walk upright on two legs.

Theory – A well-substantiated explanation of some aspect of the natural world that typically incorporates many confirmed observations, laws, and successfully verified hypotheses.

Theropod – The theropod (meaning "beast-footed") dinosaurs are a diverse group of bipedal dinosaurs. They include the largest terrestrial carnivores ever to have lived, as well as many quite small species. Theropods typically share a number of traits, including hollow, thin-walled bones and modifications of the hands and feet (three main fingers on the hand, and three main (weight-bearing) toes on the foot). Most theropods had sharp, recurved teeth for eating flesh, and claws on the ends of all of the fingers and toes. Some of these characters were lost or modified in some groups later in theropod evolution. Theropod fossils are fairly rare and often fragmentary. Fossils of small theropods are especially rare, since small bones are harder to find and are weathered away easily.

Thiagarajan, Sivasailam – The president of Workshops by Thiagi, Inc., his organization helps people improve their performance through games and simulations.

Trait – A characteristic or condition.

Transcription – The process by which messenger RNA is read from the DNA forming a gene.

Transfer RNA (tRNA) – A type of RNA that brings the amino acids to the ribosomes to make proteins. There are 20 kinds of transfer RNA molecules, one for each of the 20 main amino acids. A transfer RNA molecule has an amino acid attached to it and contains the anticodon corresponding to that amino acid in another part of its structure. In protein synthesis, each codon in the messenger RNA combines with the appropriate tRNA's anticodon, and the amino acids are arranged in order to make the protein.

Transformism – The evolutionary theory of Lamarck in which changes occur within a lineage of populations, but in which lineages do not split (i.e., no speciation occurs, at least not in the sense of the cladistic species concept) and do not go extinct.

Transition – A mutation changing one purine into the other purine, or one pyrimidine into the other pyrimidine (i.e., changes from A to G, or vice versa, or changes from C to T, or vice versa).

Transitional Fossil – A fossil or group of fossils representing a series of similar species, genera, or families, that link an older group of organisms to a younger group. Often, transitional fossils combine some traits of older, ancestral species with traits of more recent species (for instance, a series of transitional fossils documents the evolution of fully aquatic whales from terrestrial ancestors).

Translation – The process by which a protein is manufactured at a ribosome, using messenger RNA code and transfer RNA to supply the amino acids.

Transversion – A mutation changing a purine into a pyrimidine, or vice versa (i.e., changes from A or G to C or T and changes from C or T to A or G).

Trilobite – An extinct marine arthropod common from the Cambrian to Permian eras (570–245 million years ago). Trilobite fossils are abundant in rocks of this period. Trilobites were 10–675 mm long, and their flattened oval bodies were divided into three lobes by two longitudinal furrows. They had a single head shield, which bore a pair of antennae and in many species, insect-like compound eyes. This was followed by more than 20 short body segments, each with a pair of forked appendages. Many trilobites apparently burrowed in sand or mud, preying on other animals or scavenging.

Tuberculosis – An infection of the lungs, accompanied by fever and a loss of appetite, caused by the bacillus Mycobacterium tuberculosis.

Tunicate – A group of simple chordates, including sea squirts (class Ascidacea) that live attached to rocks, and the salps (class Thaliacea) that float in the sea. Tunicates are small marine animals, cylindrical, spherical, or irregular in shape, ranging from several millimeters to over 30 cm in size. They have a saclike cellulose tunic covering the body; water is drawn in through a siphon and food particles are filtered out. The free-swimming tadpole-like larvae show the major characteristics of all chordates. They subsequently undergo metamorphosis, losing their chordate features and becoming adults. One group (class Larvacea) retains their larval characteristics throughout life.

Typology – (1) The definition of classificatory groups by phenetic similarity to a "type" specimen. A species, for example, might be defined as all individuals less than x phenetic units from the species' type. (2) The theory that distinct "types" exist in nature, perhaps because they are part of some plan of nature. (See also idealism.) The type of the species is then the most important form of it, and variants around that type are noise, or "mistakes." Neo-Darwinism opposes typology because in a gene pool no one variant is any more important than any others.

U

Unequal Crossing-Over – A crossing-over in which the two chromosomes do not exchange equal lengths of DNA; one receives more than the other.

V

Van Valen, Leigh – An evolutionary biologist who came up with the model of the Red Queen – the living chess piece that Alice encounters in Through the Looking Glass who must keep running as fast as she can to stay in the same place – as a metaphor to explain evolutionary patterns. His studies involve genetics and systematics and involve a wide range of topics, including the evolution of biotas and of mammals.

Variance – A measure of how variable a set of numbers are. Technically, it is the sum of squared deviations from the mean divided by (n–1) (the number of numbers in the sample minus one). Thus, to find the variance of the set of numbers 4, 6, and 8, we first calculate the mean, which is 6. We then sum the squared deviations from the mean (4 – 6)2 + (6 – 6)2 + (8 – 6)2, which comes to 8, and divide by (n–1) (which is 2 in this case). The variance of the three numbers is 8/2 = 4. The more variable the set of numbers, the higher the variance. The variance of a set of identical numbers (such as 6, 6, and 6) is zero.

Vermeij, Geerat J. – Biologist at the Center for Population Biology of the University of California, Davis, and author of Privileged Hands: A Scientific Life. Vermeij, blind since age 3, combines autobiography and description of the evolutionary "arms race" between intertidal predator and prey species. Wider research interests include economic relationships between organisms and ecosystems and their implications for human organisms.

Vertebrates – The group (specifically, a subphylum) of animals, descended from a common ancestor, that share the derived character of an internal skeleton made of bone or cartilage.

Vestigial – Any structures that have been greatly reduced in size and function over evolutionary time to the extent that they now appear to have little or no current function.

Virulence – The disease-producing ability of a microorganism.

Virus – A kind of intracellular parasite that can replicate only inside a living cell. In its dispersal stage between host cells, a virus consists of nucleic acid that codes for a small number of genes, surrounded by a protein coat. (Less formally, according to Medawar's definition, a virus is "a piece of bad news wrapped in a protein.")

Vitamin A – A member of a chemically heterogeneous class of organic compounds that are essential, in small quantities, for life.

Von Mutius, Erika – A pediatrician and allergist, Dr. von Mutius's research interests include the epidemiology of childhood asthma and allergies with a focus on environmental predictors and gene-environment interactions.

Vrijenhoek, Robert – A senior scientist in the areas of evolutionary biology, marine biology, and conservation, Vrijenhoek studies the ecological and evolutionary consequences of genetic diversity in animals. His research efforts have focused on the evolutionary and ecological consequences of sexual and asexual reproduction in Mexican poeciliid fish (genus Poeciliopsis), as well as invertebrates in deep-sea hydrothermal vents.

W

Wake, David – A professor in the Department of Integrative Biology at the University of California, Berkeley, Dr. Wake's research emphasizes analysis of evolutionary patterns and the processes that produce them, ranging from functional morphology to evolutionary genetics and population ecology. Amphibians and reptiles are the focus of his work.

Wallace, Alfred Russel – A British naturalist and contemporary of Charles Darwin. Wallace conducted research on the Amazon River and studied the zoological differences between animal species of Asia and Australia, developing a theory of evolution similar to Darwin's.

Ward, Peter Douglas – Professor of geological sciences at the University of Washington in Seattle, where he is also an adjunct professor of zoology and astronomy. Author of several books on biodiversity and the fossil record, including Rivers in Time: The Search for Clues to Earth's Mass Extinctions and Rare Earth: Why Complex Life is Uncommon in the Universe (with Donald Brownlee). He is the principal investigator for the University of Washington's portion of the NASA Astrobiology Institute.

Wegener, Alfred – A German climatologist and geophysicist whose book, The Origins of Continents and Oceans, was the first to propose the concept of continental drift (the forerunner to the theory of plate tectonics), as well as to suggest a supercontinent called Pangaea, which Wegener suggested had fragmented into the continents as we know them today. His ideas remained controversial until the 1960s, when they became widely accepted as new evidence led to the development of the concept of plate tectonics.

White, Tim – A paleoanthropologist with the University of California, Berkeley's Laboratory for Human Evolutionary Studies, White is known for his meticulous fieldwork and analysis investigating early hominid skeletal biology, environmental context, and behavior. With an international team of colleagues, he discovered and named Ardipithecus ramidus and *Australopithecus garhi*.

Wild Type – The genotype or phenotype, out of a set of genotypes or phenotypes of a species, that is found in nature. The expression is mainly used in lab genetics to distinguish rare mutant forms of a species from the lab stock of normal individuals.

Wilford, John Noble – A New York Times reporter and winner of two Pulitzer Prizes for his national reporting of science topics, and for his work on the Challenger explosion and the aftermath. While at the Times he served as science correspondent, assistant national news editor, and director of science news.

Wilson, E. O. – A biologist and professor at Harvard University since 1955. Wilson has won two Pulitzer Prizes for his books On Human Nature and The Ants and has received numerous honors for his research and conservation efforts.

Wobble – The ability of the third base in some anticodons of tRNA to bond with more than one kind of base in the complementary position in the mRNA codon.

Woese, Carl – A molecular biologist, Dr. Woese's identification of the Archaea as a distinctive group of organisms changed the way life is classified on Earth and transformed our view of biology.

Wrangham, Richard – A primatologist, Dr. Wrangham's central interest is in the significance of chimpanzee behavior, ecology, and life-history for understanding the common ancestor between chimps and humans and subsequent human evolution.

Y

Young Earth Creationism – The belief that the universe came into being only a few 1,000 years ago. Most young Earth creationists interpret the Bible literally, including not just the special, separate creation of human beings and all other species, but the historicity of Noah's flood.

Z

Zygote – The cell formed by the fertilization of male and female gametes.

Bibliography and Webliography

1. Alley et al., (2002). *Abrupt Climate Change: Inevitable Surprises*. Committee on Abrupt Climate Change. Washington, D.C.: National Academy Press.
2. Antón, S., Spoor, F., Fellmann, C., & Swisher, C. III., (2007). *Defining Homo erectus: Size Considered*.
3. Begun, D. J., Holloway, A. K., Stevens, K., Hillier, L. W., Poh, Y. P., Hahn, M. W., Nista, P. M., Jones, C. D., Kern, A. D., & Dewey, C. N., et al., (2007). Population genomics: Whole-genome analysis of polymorphism and divergence in *Drosophila simulans*. *PLoS Biology, 5*(11), e310.
4. Benton, M. J., & Harper, D. A. T., (2009). *Introduction to Paleobiology and the Fossil Record*. Wiley-Blackwell.
5. Blount, Z., Borland, C., & Lenski, R., (2008). Historical contingency and the evolution of a key innovation in an experimental population of *Escherichia coli*. *Proc. Natl. Acad. Sci. USA, 105*, 7899–7906.
6. Bowler, P., (1985). Scientific attitudes to Darwinism in Britain and America. In: Kohn, D., (ed.), *The Darwinian Heritage* (p. 641–682). Princeton: Princeton University Press.
7. Briggs, D. E. G., (2001). In: Crowther, P. R. (ed.), *Paleobiology II*. Blackwell Scientific, Oxford.
8. Brockman, J., (2006). *Intelligent Thought: Science versus the Intelligent Design Movement*. New York: Vintage.

9. Browne, J., (1995–2002). *Charles Darwin: Voyaging; and The Power of Place* (Vols. 1, 2). London: Jonathan Cape.
10. Bryson, B., (2003). *A Short History of Nearly Everything*. New York: Broadway Books.
11. Burchfield, J., (1975, 1990). *Lord Kelvin and the Age of the Earth*. Chicago: University Of Chicago Press.
12. Burkhardt, F., (2008). *Charles Darwin: The Beagle Letters Cambridge*. Cambridge University Press.
13. Burns, K. J., Hackett, S. J., & Klein, N. K., (2002). Molecular phylogenetics of Darwin's Finches and their relatives. *Evolution, 56*, 1240–1252 (reviewed in science, 297: 1095).
14. Carroll, S. B., (2006). *Endless Forms Most Beautiful: The New Science of Evo Devo*. W.W. Norton & Co.
15. Chaisson, E., (2000). The emerging life era: A cosmological imperative. In: *Bioastronomy 99: A New Era in the Search for Life* (p. 35–42). Astronomical Society of the Pacific Conference Publication 213.
16. Coyne, J. A., & Orr, H. A., (2004). *Speciation*. Sunderland, MA: Sinauer Associates, Inc.
17. Darwin, C., (1865). *The Origin of Species by Means of Natural Selection* (6th edn.). London: John Murray.
18. Darwin, C., (1868). *The Variation of Plants and Animals under Domestication*. London: John Murray. http://www.esp.org/books/darwin/variation/facsimile/contents.htm (accessed on 27 October 2023).
19. Darwin, C., (1937). *The Voyage of the Beagle*. New York: Collier.
20. Darwin, C., (1985). In: Burkhardt, F. H., et al., (eds.), *The Correspondence of Charles Darwin*. Cambridge: Cambridge University Press. Online: http://www.darwinproject.ac.uk/ (accessed on 27 October 2023).
21. Darwin, C., (2008). In: Secord, J. A., (ed.), *Evolutionary Writings*. Oxford: Oxford University Press.
22. Darwinism and Astronomy, (1871). *Astronomical Register*, *9*, p. 116.
23. Desmond, A., & Moore, J., (1991). *Darwin, London: Michael Joseph, Penguin Group.*
24. Desmond, A., & Moore, J., (1994). *Darwin New York: Norton.*
25. DeVorkin, D., (2000). *Henry Norris Russell: Dean of American Astronomers*. Princeton: Princeton University Press.

26. Dick, S. & Strick, J., (2004). *The Living Universe: NASA and the Development of Astrobiology*. Piscataway: Rutgers University Press.
27. Eagles, G., & König, M., (2008). A model of plate kinematics in Gondwana breakup. *Geophysical Journal International, 173*, 703–717.
28. Evans, S., Jones, M., & Krause, D., (2008). A giant frog with South American affinities from the Late Cretaceous of Madagascar. *Proceedings of the US National Academy of Sciences, 105*, 2951–2956.
29. Flessa, K. W., et al., (2005). *The Geological Record of Ecological Dynamics*. National Research Council of the National Academies, National Academies Press, Washington, D.C.
30. Foote, M. J., & Miller, A. I., (2007). *Principles of Paleontology*. W. H. Freeman, New York.
31. Friedel, D. N., Snyder, L. E., Remijan, A. J., & Turner, B. E., (2005). *Detection of Interstellar Acetone Toward the Orion-KL Hot Core, ApJ, 632*, L95.
32. Gearheard, S., Matumeak, W., Angutikjuaq, I., Maslanik, J., Huntington, H. P., Leavitt, J., MatumeakKagak, D., Tigullaraq, G., & Barry, R. G., (2006). "It's not that simple": A collaborative comparison of sea ice environments, their uses, observed changes, and adaptations in Barrow, Alaska, USA, and Clyde River, Nunavut, Canada. *Ambio, 35*(4), 203–211.
33. George, J. C., Huntington, H. P., Brewster, K., Eicken, H., Norton, D. W., & Glenn, R., (2004). Observations on shorefast ice failures in Arctic Alaska and the responses of the Inupiat hunting community. *Arctic, 57*(4), 363–374.
34. Ghiselin, M., (1968). *The Triumph of the Darwinian Method*. Berkeley: University of California Press.
35. Gibbons, A., (2007). *The First Human: The Race to Discover Our Earliest Ancestors*. New York: Anchor.
36. Gould, S., (1989). *Wonderful Life: The Burgess Shale and the Nature of History*. New York: Norton.
37. Hackett, S. J., Feldheim, K., & Alvey, M., (2006). Genes and genius: The inheritance of Gregor Mendel. *DNA and Cell Biology, 25*, 655–658.
38. Hackett, S. J., Kimball, R. T., Reddy, S., Bowie, R. C. K., Braun, E. L., Braun, M. J., Chojnowski, J. L., Cox, W. A., Han, K. L., Harshman, J., Huddleston, C. J., Marks, B. D., Miglia, K. J., Moore, W. S., Sheldon,

F. H., Steadman, D. W., Witt, C. C., & Yuri, T., (2008). A phylogenomic study of birds reveals their evolutionary history. *Science, 320*, 1763–1768.

39. Hale, G., (1902). Stellar evolution in light of recent research. *Popular Science Monthly*, p. 291.
40. Hale, G., (1903). *Decennial Publications of the Yerkes Observatory III* (p. 253). Chicago.
41. Hale, G., (1908). *The Study of Stellar Evolution* (p. 3). Chicago.
42. Hamilton, L. C., Brown, B. C., & Rasmussen, R. O., (2003). West Greenland's cod-to-shrimp transition: Local dimensions of climatic change. *Arctic, 56*(3), 271–282.
43. Harington, C. R., (2008). The evolution of Arctic marine mammals. *Ecological Applications, 18*(2), S23–S40. http://www.esajournals.org/toc/ecap/18/sp2_ (accessed on 27 October 2023).
44. Harshman, J., Braun, E. L., Braun, M. J., Huddleston, C. J., Bowie, R. C. K., Chojnowski, J. L., Hackett, S., Han, K. L., Kimball, R. T., Marks, B. D., Miglia, K. J., Moore, W. S., Reddy, S., Sheldon, F. H., Steadman, D., Steppan, S., Witt, C. C., & Yuri, T., (2008). Phylogenomic evidence for multiple losses of flight in ratite birds. *Proceedings of the National Academy of Science, 105*, 13462–13467.
45. Henke, R., & Tattersall. *Handbook of Paleoanthropology* (Vol. 3, Chapter 11, pp. 1655–1693). Springer-Verlag, Berlin.
46. Hoekstra, H. E., Hirschmann, R. J., Bundey, R. A., Insel, P. A., & Crossland, J. P., (2006). A single amino acid mutation contributes to adaptive beach mouse color pattern, *Science, 313*, 101–104.
47. Hoffman, P., Kaufman, A., Halverson, G., & Schrag, D., (1998). A neoproterozoic snowball earth. *Science, 281*, 1342–1346.
48. Hollis, J. M., Lovas, F. J., Remijan, A. J., Jewell, P. R., Ilyushin, V. V., & Kleiner, I., (2006). *Detection of Acetamide (CH_3CONH_2): The Largest Interstellar Molecule with a Peptide Bond, ApJ, 643*, L25.
49. Hull, D., (1973). *Darwin and His Critics: The Reception of Darwin's Theory of Evolution by the Scientific Community*. Cambridge, Mass.: Harvard University Press.
50. Huntington, H. P., Hamilton, L. C., Nicolson, C., Brunner, R., Lynch, A., Ogilvie, A. E. J., & Voinov, A., (2007). Toward understanding the human dimensions of the rapidly changing arctic system: insights

and approaches from five HARC projects. *Regional Environmental Change*, *7*(4), 173–186.

51. Johanson, D., & Edgar, B., (2006). *From Lucy to Language: Revised, Updated, and Expanded*. New York: Simon & Schuster.
52. Katz, D. J., Edwards, T. M., Reinke, V., & Kelly, W. G., (2009). A C. elegans LSD1 demethylase contributes to germline immortality by reprogramming epigenetic memory. *Cell, 137*(2), 308–320.
53. Kerr, R., (2008). Two geologic clocks finally keeping the same time. *Science v., 320*, 434, 435.
54. Kimbel, W., Lockwood, C., Ward, C., Leakey, M., Rak, Y., & Johanson, D., (2006). Was *Australopithecus anamensis* ancestral to *A. afarensis*? A case of anagenesis in the hominin fossil record. *Journal of Human Evolution, 51*, 134–152.
55. Kimura, M., (1968). Evolutionary rate at the molecular level. *Nature, 217*, 624–626. doi: 10.1038/217624a0.
56. Krause, D., O'Connor, P., Rogers, K., Sampson, S., Buckley, G., & Rogers, R., (2006). Late Cretaceous terrestrial vertebrates from Madagascar: Implications for Latin American biogeography. *Annals of the Missouri Botanical Garden, 93*, 178–208.
57. Kuiper, K., Deino, A., Hilgen, F., Krijgsman, W., Renne, P., & Wijbrans, P., (2008). Synchronizing rock clocks of earth history. *Science, v., 320*, 500–504.
58. Lenski, R., & Travisano, M., (1994). Dynamics of adaptation and diversification: A 10,000-generation experiment with bacterial populations. *Proc. Natl. Acad. Sci. USA, 91*, 6808–6814.
59. Lenski, R., Ofria, C., Pennock, R., & Adami, C., (2003). The evolutionary origin of complex features. *Nature, 423*, 139–144.
60. Lipson, H., & Pollack, J., (2000). Automatic design and manufacture of artificial lifeforms. *Nature, 406*, 974–978.
61. Mawer, S., (2006). *Gregor Mendel: Planting the Seeds of Genetics*. Harry N. Abrams, Inc.
62. Mayr, E., & Provine, W., (1980). *The Evolutionary Synthesis: Perspectives on the Unification of Biology*. Cambridge: Harvard University Press.
63. Miller, S., & Lazcano, A., (1995). The origin of life – did it occur at high temperatures? *Journal of Molecular Evolution, 41*, 689–692.

64. Miller, S., (1953). A production of amino acids under possible primitive earth conditions. *Science, 117*, 528–529.

65. Miyakawa, S., Cleaves, H., & Miller, S., (2002). The cold origin of Life: A. Implications based on the hydrolytic stabilities of hydrogen cyanide and formamide. *Origins of Life and Evolution of the Biosphere, 32*, 195–208.

66. Moore, J., (1979). *The Post-Darwinian Controversies*. London: Cambridge University Press.

67. Moore, S. E., & Huntington, H.P., (2008). Arctic marine mammals and climate change: Impacts and resilience. *Ecological Applications, 18*(2), S157. http://www.esajournals.org/toc/ecap/18/sp2. (accessed on 27 October 2023).

68. Mullen, L. M., & Hoekstra, H. E., (2008). Natural selection along an environmental gradient: A classic cline in mouse pigmentation. *Evolution, 62*, 1555–1570.

69. Muller, G., (2007). Evo-devo: Extending the evolutionary synthesis. *Nature Reviews Genetics, 8*(1), 943–949.

70. Nichols, D., & Johnson, K., (2008). *Plants and the K-T Boundary* (p. 280). Cambridge: University Press.

71. Noonan, B., & Chippendale, P., (2006). Vicariant origin of Malagasy reptiles supports Late Cretaceous Antarctic land bridge. *American Naturalist, 168*, 730–741.

72. Noor, M. A., & Feder, J. L., (2006). Speciation genetics: Evolving approaches. *Nature Reviews Genetics, 7*(11), 851–861.

73. Numbers, R., (1977). *Creation by Natural Law*. Seattle: University of Washington Press.

74. Numbers, R., (1977). *Creation by Natural Law: Laplace's Nebular Hypothesis in American Thought. Seattle*. University of Washington Press.

75. Ortiz-Barrientos, D., Reiland, J., Hey, J., & Noor, M. A. F., (2002). Recombination and the divergence of hybridizing species. *Genetica, 116*(2–3), 167–178.

76. Pigliucci, M., (2007). Do we need an extended evolutionary synthesis? *Evolution, 61*(12), 2743–2749.

77. Price, P., (2007). Microbial life in glacial ice and implications for a cold origin of life. *FEMS Microbiology Ecology, 59*, 217–231.

78. Priscu, J., & Christner, B., (2004). Earth's icy biosphere. In: Bull, A., (ed.), *Microbial Diversity and Prospecting* (pp. 130–145). Washington, D.C.: ASM Press.

79. Remijan, A. J., Hollis, J. M., Lovas, F. J., Cordiner, M. A., Millar, T. J., Markwick-Kemper, A. J., & Jewell, P. R., (2007). *Detection of **C*_*8* *H*^*-* * **and Comparison with **C*_*8* *H **toward IRC +10 216, *^, ApJ, 2007*, L47^.

80. Remijan, A. J., Hollis, J. M., Lovas, F. J., Plusquellic, D. F., & Jewell, P. R., (2005). *Interstellar Isomers: The Importance of Bonding Energy Differences, ApJ, 632*, 333.

81. Remijan, A. J., Hollis, J. M., Lovas, F. J., Stork, W. D., Jewell, P. R., & Meier, D. S., (2008). *Detection of Interstellar Cyanoformaldehyde (CNCHO), ApJ, 675*, L85.

82. Remijan, A. J., Milam, S. N., Womack, M., Apponi, A. J., & Ziurys, L. M., et al., (2008). *The Distribution, Excitation, and Formation of Cometary Molecules: Methanol, Methyl Cyanide and Ethylene Glycol, ApJ, 689*, 613.

83. Remijan, A. J., Shiao, Y. S., Friedel, D. N., Meier, D. S., & Snyder, L. E., (2004). *A Survey of Large Biomolecules Toward Selected High Mass Star Forming Regions, ApJ, 617*, 384.

84. Rightmire, G., (1993). *The Evolution of Homo erectus: Comparative Anatomical Studies of an Extinct Human Species*. Cambridge University Press.

85. Sepkoski, D., (2009). In: Ruse, M. (eds.)., *The Paleobiological Revolution: Essays on the Growth of Modern Paleontology*. Chicago: University of Chicago Press.

86. Shermer, M., (2007). *Why Darwin Matters: The Case Against Intelligent Design*. New York: Holt.

87. Shubin, N., (2008). *Your Inner Fish: A Journey into the 3.5-Billion-Year of the Human Body*. New York: Pantheon Books.

88. Shubin, N., (2008). *Your Inner Fish: A Journey into the 3.5-Billion-Year History of the Human Body. Pantheon.*

89. Sniegowski, P., & Murphy, H., (2006). Evolvability. *Current Biology, 16*(19), R831–R834.

90. Spoor, F., Leakey, M., Gathogo, P., Brown, F., Antón, S., McDougall, I., Kiarie, C., Manthi, F., & Leakey, L., (2007). Implications of new

early Homo fossils from Ileret, east of Lake Turkana, Kenya. *Nature, 448*, 688–691.

91. Stanford, C., Allen, J., & Antón, S., (2009). *Introduction to Biological Anthropology* (2nd edn.) Pearson Prentice Hall.
92. Steiner, C. C., Rompler, H., Boettger, L. M., Schoneberg, T., & Hoekstra, H. E., (2009). The genetic basis of phenotypic convergence in beach mice: Similar pigment patterns but different genes. *Molecular Biology and Evolution, 26*, 35–45.
93. Steiner, C. C., Weber, J. N., & Hoekstra, H. E., (2007). Adaptive variation in beach mice produced by interacting pigmentation genes. *PLoS Biol., 5*, e219.
94. Stringer, C., & Andrews, P., (2005). *The Complete World of Human Evolution*. London: Thames & Hudson.
95. Sulloway, F., (1982). Darwin and his finches: The evolution of a legend. *Journal of the History of Biology, 15*, 1–53.
96. Sulloway, F., (1982). Darwin's conversion: The beagle voyage and its aftermath. *Journal of the History of Biology, 15*, 325–396.
97. Trinks, H., Schroder, W., & Biebricher, C., (2005). Ice and the origins of life. *Origins of Life and Evolution of Biospheres, 35*, 429–445.
98. Walker, G., (2004). *Snowball Earth*. London: Bloomsbury Publishing PLC.
99. Wallace, A., (1881). *Island Life or the Phenomena and Causes of Insular Faunas and Floras, Including a Revision and Attempted Solution of the Problem of Geological Climates*. New York: Harper.
100. Weiner, J., (1994). *The Beak of the Finch: A Story of Evolution in Our Time*. Alfred A. Knopf, New York.
101. Weiner, J., (1994). *The Beak of the Finch: A Story of Evolution in Our Time*. New York: Knopf.
102. Weiner, J., (1995). *The Beak of the Finch*. New York: Vintage Books.
103. Weiss, K., & Fullerton, S., (2005). Racing around, getting nowhere. *Evolutionary Anthropology, 14*, 165–169.
104. Weiss, K., (2001). We hold these truths to be self-evident. *Evolutionary Anthropology, 10*, 199–203.
105. Weiss, K., (2004). The smallest grain in the balance. *Evolutionary Anthropology, 13*, 122–126.

106. Weiss, K., (2007). The scopes trial. *Evolutionary Anthropology, 16*, 126–131.

107. Weiss, K., (2008). The good, the bad, and the Ugli. *Evolutionary Anthropology, 17*, 129–134.

108. Widicus Weaver, S. L., Remijan, A. J., McMahon, R. J., & McCall, B. J., (2007). *Investigating Benzene Derivatives Toward Protoplanetary Nebulae: **A Search for */*o*/*-Benzyne (*/*o*/*-C*_*6* *H*_*4* *) and Phenyl (C*_*6* *H*_*5* *) toward CRL 618, ApJ, 671*, L153.

109. Woods, R., Schneider, D., Winkworth, C., Riley, M., & Lenski, R., (2006). Tests of parallel molecular evolution in a long-term experiment with *Escherichia coli. Proc. Natl. Acad. Sci. USA, 103*, 9107–9112.

110. Wright, H., (1994). *Explorer of the Universe: A Biography of George Ellery Hale*. Reprinted, American Institute of Physics.

111. Young, C., (1884). Pending problems in astronomy. In: *The Observatory* (Vol. 7, p. 283).

INDEX

C

D

T

U

V

W